AF454303

Cognitive Aging

Cognitive Aging

Special Issue Editor

Myra Fernandes

MDPI • Basel • Beijing • Wuhan • Barcelona • Belgrade • Manchester • Tokyo • Cluj • Tianjin

Special Issue Editor
Myra Fernandes
University of Waterloo
Canada

Editorial Office
MDPI
St. Alban-Anlage 66
4052 Basel, Switzerland

This is a reprint of articles from the Special Issue published online in the open access journal *Brain Sciences* (ISSN 2076-3425) (available at: https://www.mdpi.com/journal/brainsci/special_issues/cognitive_aging).

For citation purposes, cite each article independently as indicated on the article page online and as indicated below:

LastName, A.A.; LastName, B.B.; LastName, C.C. Article Title. *Journal Name* **Year**, *Article Number*, Page Range.

ISBN 978-3-03936-136-6 (Hbk)
ISBN 978-3-03936-137-3 (PDF)

Cover image courtesy of pixabay.com.

Contents

About the Special Issue Editor

Myra Fernandes is a Professor in Cognitive Neuroscience at the University of Waterloo. Her research identifies cognitive processes and key brain structures supporting memory. She uses procedures such as fMRI and behavioral testing in young adults, aging individuals, and in those with a past head injury or concussion. Dr. Fernandes was awarded the Canadian Psychological Association's President's New Researcher Award, the Ontario Ministry of Research and Innovation's Research Excellence Award, and the Women in Cognitive Science Canada Mentorship Award. She holds numerous editorial board positions and is a fellow of the Association for Psychological Science, and Canadian Society for Brain Behaviour and Cognitive Sciences. She was a Co-Chair of the Natural Sciences and Engineering Research Council of Canada's grant review panel for Biological Systems. Dr. Fernandes is also a past chair of NSERC's Scholarships and Fellowships committee, as well as being the past chair of the Ontario Graduate Scholarships panel. Dr. Fernandes has published numerous peer-reviewed articles, multiple book chapters, and books. Her research is frequently featured in newspapers, magazines, radio interviews, blogs, and science documentaries worldwide, from journalists in the areas of Psychology, Law, Medical Sciences, Business, Biology, and Technology. Dr. Fernandes holds a BSc degree in Psychology and Biology from the University of Waterloo (1995), as well as a Masters (1996) and PhD (2001) from the University of Toronto, in Cognitive Neuropsychology.

Preface to "Cognitive Aging"

Given the global demographic shift towards an aging population, there is a pressing need to understand how aging affects cognition. This Special Issue highlighted research investigating the fundamental processes and mechanisms underlying the neural and cognitive changes associated with aging.

Myra Fernandes
Special Issue Editor

Review

A Homeostatic Model of Subjective Cognitive Decline

Akiko Mizuno [1,*], Maria Ly [1,2] and Howard J. Aizenstein [1,3]

[1] Department of Psychiatry, University of Pittsburgh, Pittsburgh, PA 15213, USA;
    Ly.Maria@medstudent.pitt.edu (M.L.); aizen@pitt.edu (H.J.A.)
[2] Department of Neuroscience, University of Pittsburgh, Pittsburgh, PA 15213, USA
[3] Department of Bioengineering, University of Pittsburgh, Pittsburgh, PA 15213, USA
*   Correspondence: akm82@pitt.edu

Received: 31 October 2018; Accepted: 17 December 2018; Published: 19 December 2018

Abstract: Subjective Cognitive Decline (SCD) is possibly one of the earliest detectable signs of dementia, but we do not know which mental processes lead to elevated concern. In this narrative review, we will summarize the previous literature on the biomarkers and functional neuroanatomy of SCD. In order to extend upon the prevailing theory of SCD, compensatory hyperactivation, we will introduce a new model: the breakdown of homeostasis in the prediction error minimization system. A cognitive prediction error is a discrepancy between an implicit cognitive prediction and the corresponding outcome. Experiencing frequent prediction errors may be a primary source of elevated subjective concern. Our homeostasis breakdown model provides an explanation for the progression from both normal cognition to SCD and from SCD to advanced dementia stages.

Keywords: subjective cognitive decline; preclinical dementia; fMRI; compensation

---

1. Introduction

Subjective Cognitive Decline (SCD) refers to an individual's perception that their cognitive performance has declined, despite having no significant objective cognitive impairment. SCD may reflect one of the earliest signs of dementia, as it is a risk factor for developing mild cognitive impairment (MCI) and Alzheimer's disease (AD) [1,2]. However, SCD is quite understudied—which mental processes lead to SCD and the neural basis of SCD are yet to be understood. Here, we will provide a narrative review of the current literature for biomarkers, and the functional neuroanatomy associated with SCD. Then, we will propose a new model that integrates existing findings in SCD into a new neural system dysfunction model, which involves heightened prediction of error-signaling and homeostatic breakdown.

2. AD Biomarkers (Neurodegenerative Factors) and SCD

In order to investigate whether SCD represents a pre-clinical state of AD, the relationships between AD biomarkers in individuals with SCD were examined. In the traditional AD biomarker cascade, amyloid (Aβ) accumulation occurs prior to neurodegeneration and cognitive decline [3]. The most promising evidence that SCD precedes MCI and AD is that Aβ deposition is associated with SCD symptom severity, but not with objective memory performance [4–6]. Vogel and colleagues [7] found that amyloid status predicted future cognitive decline (on average, four years) among individuals with SCD. However, a larger longitudinal study [8] concluded that the Aβ-status by itself does not predict the progression of AD over a relatively brief time span of 2.5 years. More longitudinal studies are necessary to better understand the relationship between Aβ and objective cognitive decline in SCD. In addition to the accumulation of amyloid plaques, AD is also associated with neurofibrillary tangles composed of tau protein. Tau accumulation is believed to be more closely related to neurodegeneration and cognitive decline in AD, as compared with amyloid [9]. Tau has been shown to be more associated with SCD than with non-amnestic MCI [10]. Thus, like Aβ, the tau markers support the role SCD as an AD risk group.

Brain atrophy and white matter hyperintensities have been also reported in SCD, as seen in the early pre-clinical stages of AD progression. Several studies consistently reported cortical volume loss and thinning in the medial temporal regions in those with SCD [11–15], indicating the decreased structural integrity of the memory system. Longitudinal observations also reported the association between atrophy and future cognitive decline in SCD [16,17]. Whole-brain analysis by Verfaillie and colleagues [17] suggested that a steeper decline in cognition was not only associated with a thinner cortex of the temporal region but also the frontal and occipital cortices in SCD. Increased amounts of white matter hyperintensities in widespread regions have also been reported in SCD [18,19]. These studies provide support for the idea that SCD may be an early transitional stage prior to the onset of dementia (i.e., MCI and AD), especially as seen with the perturbations in the memory systems.

3. Functional Neuroanatomy and Compensation Theory in SCD

The neural basis of elevated subjective concern for cognitive decline among older individuals with normal cognition is the least investigated, but it represents a growing area of research (Table 1). Four functional magnetic resonance imaging (fMRI) studies investigated brain activation during memory-related tasks [20–23]. Most of these studies did not find group differences between participants with and without SCD in the behavioral performance of the task, but observed different functional brain activation patterns. The first study by Rodda and colleagues [23] measured brain activity while participants were encoding a list of semantically related words, which were later tested through a recognition paradigm. Whole-brain analysis demonstrated increased activation in the lateral part of the prefrontal cortex (PFC) in those with SCD. The level of the PFC activation was positively correlated with task performance. The authors interpreted that increased PFC activation served as neural compensation for the decreasing function of the primary hippocampal memory system, as indicated in previous structural imaging studies (summarized in the previous section). To test this compensation hypothesis, Erk et al. (2011) [20] investigated activation in the hippocampus and the PFC during memory encoding of faces and associated occupations, through a region-of-interest approach. Their results demonstrated decreased activation in the hippocampus, and increased activation in the dorsolateral PFC (dlPFC). Task performance was positively correlated with dlPFC activation only in the SCD group, which provided support for the compensation hypothesis in SCD.

Table 1. Summary of fMRI studies in SCD. Most of these studies demonstrate a pattern of regional hypoactivation, associated with hyperactivation elsewhere. This hyperactivation has been interpreted by authors as a compensatory response to the hypoactivation. Abbreviations: SCD: subjective cognitive decline, PFC: prefrontal cortex, BA: Brodmann area, DLPFC: dorsolateral prefrontal cortex, ROI: region-of-interest, SPL: superior parietal lobe, PCC: posterior cingulate cortex, DMN: default mode network, VMPFC: ventromedial prefrontal cortex, ACC: anterior cingulate cortex, MFG: middle frontal gyrus, WM: working memory.

Reference	fMRI Task	Participants	Hyperactivation (SCD > Control) or Positive Correlation with SCD Symptoms	Hypoactivation (Control > SCD) or Negative Correlation with SCD Symptoms	Behavioral Performance
Rodda et al. (2009) [23]	Memory encoding	10 memory clinic SCD vs. 10 controls (age: 64.2 vs. 68.0)	L PFC (BA6/9/44/46)		(1) No group differences in behavioral performance (2) Positive correlation between PFC activation and recognition performance in both groups.
Erk et al. (2011) [20]	Memory (encoding, recall, recognition) and working memory (n-back)	19 memory clinic SCD vs. 20 controls (age: 68.4 vs. 66.8)	R DLPFC during recall (ROI analysis)	Hippocampus during recall (ROI analysis)	(1) No group differences in behavioral performance (2) Positive correlation between DLPFC activation and recognition performance in SCD. (3) Positive correlation between hippocampal activation and recognition performance in controls
Rodda et al. (2011) [24]	Divided attention	11 memory clinic SCD vs. 10 controls (age: 64.6 vs. 68.0)	L medial temporal, bilateral thalamus, PCC, caudate		(1) No group differences in behavioral performance
Dumas et al. (2013) [25]	Working memory (n-back)	Postmenopausal women: 12 cognitive complainers vs. 11 controls (age = 56.8 vs. 57.1)	MFG (BA10/9), ACC (BA24/32), insula (BA 13), precuneus (increased activation as WM demand increased)	Caudate	(1) No group differences in behavioral performance
Hu et al. (2017) [22]	Future-oriented decision making	20 memory clinic SCD vs. 24 controls (age: 68.3 vs. 66.49)		Medial frontal polar cortex, ACC, insula	(2) SCD showed reduced future-oriented choices (3) Positive correlation between hippocampal activation (ROI analysis) and future-oriented choice in only the control.
Hayes et al. (2017) [21]	Memory: successful vs. unsuccessful encoding	23 SCD vs. 41 controls (age: 68.6 vs 67.5) 21 out of 23 were memory clinic SCD	(1) Negative subsequent memory effect in the occipital lobe, SPL, PCC (2) More complaints, more negative subsequence memory effects in DMN (PCC, precuneus, VMPFC)		(1) No group differences in behavioral performance

Hu et al. (2017) [22] utilized a task that emulated memory processes that were relevant to activities in daily life (future-oriented choice tasks). Brain activation was measured, while participants were required to select an immediate or delayed reward regarding a personally relevant episodic future event. Successful selection of the future-oriented choice (i.e., delayed reward) over the immediate reward requires the crucial involvement of episodic memory and valuation systems [26]. Unlike other studies, their study is the only one which observed group differences in task performance. The SCD group showed reduced preference for future-oriented choice, which was previously demonstrated in those with MCI [27]. A priori region-of-interest analysis showed that only participants in the control group showed an association between greater hippocampal activation and more future-oriented choices. The whole-brain analyses found reduced activation in medial frontal regions (medial frontal pole and anterior cingulate cortex (ACC)) and the insula in the SCD group, suggesting diminished valuation functioning. The authors suggested that reduced involvement in the episodic memory and valuation system in SCD during the decision-making process may reflect the attenuating attention and subjective evaluation system.

To address the possibility that increased PFC activation reflects a general cognitive processes rather than only memory encoding, Hayes et al. (2017) [21] used an event-related design to compare brain activation for high-confidence successful recall versus failed recall. The SCD group showed an increased activation for failed recall (i.e., negative subsequent memory) in the posterior areas (occipital, superior parietal, and precuneus). They also regressed activation on a continuous SCD symptom severity to identify the neural correlates of SCD, by combining participants in both groups. Participants with more severe SCD symptoms showed increased activation for failed recall in both the frontal and posterior nodes of the default mode network, which normally suppress its activity during cognitive tasks. They concluded that individuals with SCD rely on the altered neural system for successful memory encoding to maintain normal cognitive function.

Cognitive concerns in SCD mainly reflect the perceived decline of memory function, but the other domains of cognitive function may also contribute to elevated concerns [28]. There are two fMRI studies with non-episodic memory tasks. Dumas and colleagues [25] measured brain activation in SCD by using the n-back working memory task. Although this study was limited to females (i.e., a comparison between those with and without cognitive concerns among postmenopausal women without hormone therapy), their study is the first to report that activation increased with increasing cognitive load/effort among those with SCD. These effects were found in the extended working memory system, including middle frontal (BA 9/10), anterior cingulate cortex (BA 24/32), and the precuneus (BA 13). Both groups showed the same levels of behavioral performance. Another study by Rodda et al. (2011) [24] investigated brain activation during a divided attention task, where participants with SCD were required to respond to target stimuli while processing sequences of both visual and auditory information. Behavioral performance did not show group differences, but the SCD group demonstrated increased activation in two medial posterior regions: one in the cerebellum, and another in the thalamus, extending to the posterior cingulate cortex and the medial temporal lobe (hippocampus and parahippocampus). These two studies are consistent with the idea that early functional changes (i.e., increased activation) in executive function may manifest in SCD, despite the lack of impairment in behavioral performance or in the neuropsychological tests.

In summary, previous fMRI studies, along with the structural imaging studies (summarized in the previous section) have suggested three neural phenomena in SCD: (1) loss of integration of the memory system, (2) compensatory hyperactivation in the prefrontal cortex or the use of other alternative neural resources to maintain normal performance, and (3) decreased prefrontal activation for subtle yet declining higher-order cognitive functions. In other words, the direction of neural activation (increased or decreased) observed in SCD depends on whether the expected level of performance can be maintained. However, it is important to note that the sample sizes and statistical power in these studies were relatively low; likewise, inconsistencies between studies may have been partially due to sampling error. To extend our understanding of the functional neuroanatomy of SCD, more mechanistic and cohesive frameworks that can provide explanations for dysfunctional processes are necessary, regardless the level of task performance or cognitive domain. Furthermore, it is not yet understood which specific cognitive processes rely upon compensation, and how these processes are directly associated with SCD symptoms.

4. Current Theories of the Neural Basis of SCD

Compensatory hyperactivation of the prefrontal region is currently the most popular theory of the neural basis of SCD. Another theory is brain reserve [29], which proposes a structural basis for functional compensatory capacity. Alternatively, the dedifferentiation [30], a loss of specialization of neural function resulting in diffuse brain activations, is a theory which may explain hyperactivation in the prefrontal cortex. All of these versions of compensatory hyperactivation describe only the temporal transition from the pre-SCD state to the SCD state, and do not describe the dynamics of post-SCD neurodegeneration. Further, they do not describe how hyperactivation may contribute to post-SCD decline via harmful biological effects on the neural system, such as neurotoxicity or excitotoxicity. Here, we introduce homeostasis breakdown, a new mechanistic framework for comprehensive temporal dynamics in SCD and progression to AD.

5. Background of the Prediction Error Theory

Our brain functions as a statistical optimization engine that constantly makes implicit predictions of sensory inputs [31,32]. That is, rather than passively receiving sensory information, it is actively making inferences. These inferences are propagated as predicted expectations to heteromodal association areas and the PFC. These expectations are compared with the current environment, and a behavior is chosen. Moreover, the difference between predicted and observed behavior is used as learning signal to adapt for better performance in the future. This constant process of comparing internally generated predictions with external reality is called "predictive coding" [33]. This is why we think of the brain as learning and adapting across all behaviors.

Prediction errors refer to the mismatch between the internally generated prediction and the external reality. The most prominent brain region that responds to such errors is the dorsal anterior cingulate cortex (dACC). Both animal and human studies have demonstrated the increased activity in dACC to response to prediction errors [34,35]. Similar terms for prediction errors are conflict [34,36] and free-energy [37]. Although there is a general consensus that dACC mediates error-related signals; neuroscientists have different opinions about the specific processes in the dACC and the primary goal of the function [38]. Nonetheless (regardless of the diverse terminology and theories of dACC function), the minimization of prediction error is a core organizing principal for computational function at the local neural circuit. This error minimization optimizes our internal predictions, which facilitates successful goal-directed behaviors and survival.

6. Prediction Error and SCD Symptoms

Suppose a man who is very experienced with a computer notices that he is making more typing errors. If he experiences subtle yet frequent errors between his prediction ("I thought I typed out 'experience'") and actual outcome ("I accidentally typed 'exprience'"), his level of SCD symptoms may rise. This type of error signal raises the activation in the dACC, resulting in varying levels of awareness. We believe that an uncharacteristic accumulation of implicit errors may gradually lead to more effort being required to maintain cognitive performance, and then to having more explicit levels of error awareness. Another example that may highlight the relationship between error monitoring and the experience of cognitive tasks being more effortful in SCD would be the following: an individual may be accustomed to finishing the "New York Times" crossword puzzle in 20 minutes (their prediction). However, if that individual finds that they are now needing 30 minutes or more to complete the puzzle, this reflects both a prediction error and the experience of increased effort. Such awareness of errors could occur not only in memory, but rather across multiple cognitive domains, including attention, task switching, language, and mathematical operations.

Awareness of one's internal cognitive system is called metacognition. According to Nelson (1990) [39], metacognition has two primary operations: monitoring and control. Monitoring refers to the introspection of incoming sensory information and one's own performance, whereas control refers to an allocation of an action (i.e., self-regulation). These two operations are independent, but reciprocally interacted. Both the prediction error [31,32] and conflict monitoring [34,36] frameworks provide mechanistic models for the monitoring of errors or conflicts. These models, however, differ in the level of information processing that they are meant to explain. Prediction error refers to the process that can occur throughout the cortical network. Prediction errors provide a signal that our internal model need to be updated, and the signal is generated by distributed processes of our incoming sensory information [40]. These local prediction errors influence the local Hebbian learning model [41]. In this way, correct predictions are strengthened, and incorrect predictions are weakened [42]. On the other hand, conflict-monitoring framework refers to how the more extended controlled yet implicit cognitive process integrates generated error signals, such as prepotent response suppression. Unlike prediction error, which is a general network learning signal, conflict monitoring refers to the specific monitoring and control functions in the dACC. The prediction error [31,32] process may specifically infer to the earlier operation of monitoring, whereas conflict-monitoring [34,36] may associate with both monitoring and control operations, suggested in Nelson's framework [39].

In the framework of metacognition, SCD can be interpreted as the impairment of both monitoring and the early stages of control. The accumulated subjective experience of prediction errors and perceived increase in an effort to complete tasks may eventually lead to the elevated self-awareness of cognitive decline, as reflected in SCD. As these individuals develop frank cognitive decline, conflict-monitoring processes may then decrease, leading to an opposite pattern of decreased self-awareness of cognitive functioning. Experimental tasks that are sensitive to the suppression of the inappropriate responses to implicit prediction error may be able to capture an earlier objective process of decline in SCD. Furthermore, neuroimaging studies with these tasks will provide detailed neural mechanisms of dysfunctional metacognition in SCD.

7. Prediction Error and SCD Characteristics

Previous studies suggest that the cognitive implications of SCD symptoms depend on the level of education achievement [28]. Since higher level of education is considered to be a marker of cognitive reserve, Stern [43,44] postulated that cognitive reserve provides resilience to neurodegeneration. Levels of cognitive reserve may represent the sensitivity to prediction errors, and the utility of the error signals. Individuals with high cognitive reserve may be highly sensitive to prediction errors, and interpret them as important learning signals to update the internal model. These individuals constitute a lifestyle of using these learning signals more frequently and effectively to make higher achievements, resulting in having a high cognitive reserve.

Individuals with SCD are often highly anxious, and characterized by their tendency to worry [1], usually expressed as high neuroticism in the Big Five personality trait model [45]. It has been demonstrated that individuals with high neuroticism are highly sensitive to prediction errors [46]. In the course of progression of neurodegeneration, these individuals may start noticing prediction errors earlier than those with low neuroticism. The frequent experience of errors may not only raise an awareness, but may also elicit concern. Individual with high neuroticism may then interpret the perceived errors as important learning signals, resulting in symptoms of SCD.

The high prevalence of depressive symptoms is another characteristic that has been reported consistently in SCD [1]. Neuroticism is also highly associated with depression; however, the symptoms associated with neuroticism and depression may relate to different aspects of SCD. Neuroticism may serve as a predictor of how an individual may interpret prediction errors, whereas depressive symptoms may reflect the affective response to an individual's interpretation of their prediction errors.

Depressive symptoms in SCD, therefore, may be translated as a negative affective response (i.e., sad feeling) to frequently experiencing errors (i.e., the "monitoring" component of metacognition), leading to the persistence of depressive moods over time [47]. Depressive symptoms may also be a form of adjustment disorder, where an individual may have an emotional reaction to their new experience of difficulty in both internal prediction and performance (i.e., the "control" component of metacognition).

8. Homeostasis Breakdown

Homeostasis—or homeostatic regulation—is the ability to maintain stability and equilibrium of the system. As a classical example, the stability of our body temperature is a consequence of homeostatic processes that coordinate the activity of muscles, blood vessels, and sweat glands. When a cold environment decreases body temperature, the hypothalamus releases a signal to the skeletal muscles, promoting shivering as a mechanism of thermogenesis and a signal to the blood vessels to increase resistance of blood flow (i.e., vasoconstriction). Both of these responses minimize heat loss, helping to maintain body temperature.

Prior work by Li et al. has suggested that homeostatic dysregulation in multiple systems occur in aging, and may also serve as a key contributor to the biological mechanisms of aging [48]. While the authors have provided evidence in the systems of lipids, immune function, oxygen transport, liver functioning, vitamin levels, and electrolyte levels, they suggest that homeostatic dysregulation is not limited to these systems, and may occur in other systems in aging.

Thus, we propose that compensatory hyperactivation may represent a homeostatic process that serves to maintain the stability of cognition in a changing neurobiological environment. Homeostasis in the context of cognition serves to maintain cognitive functioning, despite the presence of neurodegeneration. Neurodegeneration may lead to prediction errors and corresponding SCD symptoms, much like the body temperature falling just enough to cause a sensation of coldness. Finally, the compensatory hyperactivation is one of the main homeostatic processes that we are currently aware of in cognition, and this is analogous to the onset of vasoconstriction of blood vessels to prevent hypothermia.

However, homeostatic processes can have negative side effects. For example, extreme vasoconstriction for an extended period of time can lead to vascular cell loss. Similarly, compensatory hyperactivation may lead to glutaminergic excitotoxicity, which may lead to neuronal death [49] or the production of Aβ [50]. Thus, although homeostasis can slow the onset of cognitive decline, this may come at the cost of negative side effects that weaken the core cognitive infrastructure. This may explain why individuals with SCD tend to experience a relatively rapid decline into AD [51].

This homeostatic model of SCD represents an extension of the prevailing compensatory hyperactivation model, as it may provide a mechanistic explanation for differing levels of neural activity. The compensation model is unclear as to what entity "drives" the change in activation levels, whereas the homeostatic model provides a dynamic control system that maintains cognitive functioning that is dependent on prediction error and conflict monitoring.

9. Future Directions

Clinicians do not yet have a standard intervention protocol for individuals with SCD. If the neural basis for SCD were better understood, an effective intervention may be developed. A recent meta-analysis of experimental interventions for SCD suggested that cognitive restructuring therapies may improve metacognition (i.e., alleviate self-perceived cognitive challenges) [52], indicating that SCD could be a modifiable risk factor of dementia. Alleviating SCD symptoms, along with associated psychological distress may slow neurodegeneration, such as atrophy and Aβ accumulation, by reducing hyperactivation.

More studies investigating the markers of neurotoxicity in SCD are necessary to provide basic evidence for psychotherapeutic interventions in the earliest stages of dementia. In this narrative review, we have proposed a homeostatic prediction error model for understanding how the progression of neural system dysfunction can manifest as SCD. fMRI studies can help in validating that increases in error prediction, and conflict monitoring is central to the subjective perception of cognitive impairment. Further research is also necessary to identify the working elements (neural control system, sensors, set point, etc.) of the homeostatic prediction error model. Moreover, fMRI studies coupled with behavioral interventions can be used as an outcome measure. This would help in the development and adaptation of behavioral interventions targeting SCD, and potentially mitigating the accelerated neurodegeneration that may be associated with hyperactivated stressed system that is out of homeostatic balance.

10. Conclusions

In common scientific practice, the term *subjective* may generally be disfavored, because it connotes a lack of objectivity, as the self-assessments that are used to diagnose SCD, presumably include individual biases. However, the presence of SCD symptoms appears to contain valuable information regarding cognitive decline over time and underlying neurophysiological pathologies. More studies and theoretical frameworks that can comprehensively explain temporal dynamics, including the positive and negative by-products of compensatory hyperactivation, are necessary. In this review, we proposed that prediction error, a metacognitive process, potentially leads to SCD symptoms. We also introduced homeostatic breakdown as a new framework that incorporates and integrates the current findings with the new prediction error perspective, to describe the cascading effect of neurodegeneration and cognitive decline in SCD. This framework has the potential to motivate new standard therapies for SCD that focus on alleviating not only the subjective symptoms, but also slow the progression of dementia, due to neurotoxicity from compensatory hyperactivation.

Funding: This study was supported by grants T32 MH019986, T32 AG021885, P50 AG005133, P01 AG025204, and R37 AG025516 from the National Institute of Health.

Acknowledgments: We thank the administrative staff of the Geriatric Psychiatry Neuroimaging lab for assistance with various logistics in the preparation of this manuscript.

Conflicts of Interest: The authors declare no conflicts of interest.

References

1. Jessen, F.; Amariglio, R.E.; van Boxtel, M.; Breteler, M.; Ceccaldi, M.; Chetelat, G.; Dubois, B.; Dufouil, C.; Ellis, K.A.; van der Flier, W.M.; et al. A conceptual framework for research on subjective cognitive decline in preclinical Alzheimer's disease. *Alzheimers Dement.* **2014**, *10*, 844–852. [CrossRef] [PubMed]
2. Mitchell, A.J.; Beaumont, H.; Ferguson, D.; Yadegarfar, M.; Stubbs, B. Risk of dementia and mild cognitive impairment in older people with subjective memory complaints: Meta-analysis. *Acta Psychiatr. Scand.* **2014**, *130*, 439–451. [CrossRef] [PubMed]
3. Jack, C.R., Jr.; Knopman, D.S.; Jagust, W.J.; Shaw, L.M.; Aisen, P.S.; Weiner, M.W.; Petersen, R.C.; Trojanowski, J.Q. Hypothetical model of dynamic biomarkers of the Alzheimer's pathological cascade. *Lancet Neurol.* **2010**, *9*, 119–128. [CrossRef]
4. Amariglio, R.E.; Becker, J.A.; Carmasin, J.; Wadsworth, L.P.; Lorius, N.; Sullivan, C.; Maye, J.E.; Gidicsin, C.; Pepin, L.C.; Sperling, R.A.; et al. Subjective cognitive complaints and amyloid burden in cognitively normal older individuals. *Neuropsychologia* **2012**, *50*, 2880–2886. [CrossRef] [PubMed]
5. Perrotin, A.; Mormino, E.C.; Madison, C.M.; Hayenga, A.O.; Jagust, W.J. Subjective cognition and amyloid deposition imaging: A Pittsburgh Compound B positron emission tomography study in normal elderly individuals. *Arch. Neurol.* **2012**, *69*, 223–229. [CrossRef] [PubMed]
6. Snitz, B.E.; Weissfeld, L.A.; Cohen, A.D.; Lopez, O.L.; Nebes, R.D.; Aizenstein, H.J.; McDade, E.; Price, J.C.; Mathis, C.A.; Klunk, W.E. Subjective Cognitive Complaints, Personality and Brain Amyloid-beta in Cognitively Normal Older Adults. *Am. J. Geriatr. Psychiatry* **2015**, *23*, 985–993. [CrossRef] [PubMed]
7. Vogel, J.W.; Varga Dolezalova, M.; La Joie, R.; Marks, S.M.; Schwimmer, H.D.; Landau, S.M.; Jagust, W.J. Subjective cognitive decline and beta-amyloid burden predict cognitive change in healthy elderly. *Neurology* **2017**, *89*, 2002–2009. [CrossRef]

8. Dubois, B.; Epelbaum, S.; Nyasse, F.; Bakardjian, H.; Gagliardi, G.; Uspenskaya, O.; Houot, M.; Lista, S.; Cacciamani, F.; Potier, M.C.; et al. Cognitive and neuroimaging features and brain beta-amyloidosis in individuals at risk of Alzheimer's disease (INSIGHT-preAD): A longitudinal observational study. *Lancet Neurol.* **2018**, *17*, 335–346. [CrossRef]

9. Bejanin, A.; Schonhaut, D.R.; La Joie, R.; Kramer, J.H.; Baker, S.L.; Sosa, N.; Ayakta, N.; Cantwell, A.; Janabi, M.; Lauriola, M.; et al. Tau pathology and neurodegeneration contribute to cognitive impairment in Alzheimer's disease. *Brain* **2017**, *140*, 3286–3300. [CrossRef]

10. Eliassen, C.F.; Reinvang, I.; Selnes, P.; Grambaite, R.; Fladby, T.; Hessen, E. Biomarkers in subtypes of mild cognitive impairment and subjective cognitive decline. *Brain Behav.* **2017**, *7*, e00776. [CrossRef]

11. Hu, X.; Teunissen, C.; Spottke, A.; Heneka, M.T.; Duzel, E.; Peters, O.; Li, S.; Priller, J.; Burger, K.; Teipel, S.; et al. Smaller medial temporal lobe volumes in individuals with subjective cognitive decline and biomarker evidence of Alzheimer's disease-Data from three memory clinic studies. *Alzheimers Dement.* 2018.

12. Jessen, F.; Feyen, L.; Freymann, K.; Tepest, R.; Maier, W.; Heun, R.; Schild, H.H.; Scheef, L. Volume reduction of the entorhinal cortex in subjective memory impairment. *Neurobiol. Aging* **2006**, *27*, 1751–1756. [CrossRef] [PubMed]

13. Saykin, A.J.; Wishart, H.A.; Rabin, L.A.; Santulli, R.B.; Flashman, L.A.; West, J.D.; McHugh, T.L.; Mamourian, A.C. Older adults with cognitive complaints show brain atrophy similar to that of amnestic MCI. *Neurology* **2006**, *67*, 834–842. [CrossRef] [PubMed]

14. Striepens, N.; Scheef, L.; Wind, A.; Popp, J.; Spottke, A.; Cooper-Mahkorn, D.; Suliman, H.; Wagner, M.; Schild, H.H.; Jessen, F. Volume loss of the medial temporal lobe structures in subjective memory impairment. *Dement. Geriatr. Cognit. Disord.* **2010**, *29*, 75–81. [CrossRef] [PubMed]

15. van der Flier, W.M.; van Buchem, M.A.; Weverling-Rijnsburger, A.W.; Mutsaers, E.R.; Bollen, E.L.; Admiraal-Behloul, F.; Westendorp, R.G.; Middelkoop, H.A. Memory complaints in patients with normal cognition are associated with smaller hippocampal volumes. *J. Neurol.* **2004**, *251*, 671–675. [CrossRef] [PubMed]

16. Stewart, R.; Godin, O.; Crivello, F.; Maillard, P.; Mazoyer, B.; Tzourio, C.; Dufouil, C. Longitudinal neuroimaging correlates of subjective memory impairment: 4-year prospective community study. *Br. J. Psychiatry* **2011**, *198*, 199–205. [CrossRef] [PubMed]

17. Verfaillie, S.C.J.; Slot, R.E.; Tijms, B.M.; Bouwman, F.; Benedictus, M.R.; Overbeek, J.M.; Koene, T.; Vrenken, H.; Scheltens, P.; Barkhof, F.; et al. Thinner cortex in patients with subjective cognitive decline is associated with steeper decline of memory. *Neurobiol. Aging* **2018**, *61*, 238–244. [CrossRef] [PubMed]

18. Li, X.Y.; Tang, Z.C.; Sun, Y.; Tian, J.; Liu, Z.Y.; Han, Y. White matter degeneration in subjective cognitive decline: A diffusion tensor imaging study. *Oncotarget* **2016**, *7*, 54405–54414. [CrossRef]

19. Selnes, P.; Fjell, A.M.; Gjerstad, L.; Bjornerud, A.; Wallin, A.; Due-Tonnessen, P.; Grambaite, R.; Stenset, V.; Fladby, T. White matter imaging changes in subjective and mild cognitive impairment. *Alzheimers Dement.* **2012**, *8*, S112–S121. [CrossRef]

20. Erk, S.; Spottke, A.; Meisen, A.; Wagner, M.; Walter, H.; Jessen, F. Evidence of neuronal compensation during episodic memory in subjective memory impairment. *Arch. Gen. Psychiatry* **2011**, *68*, 845–852. [CrossRef]

21. Hayes, J.M.; Tang, L.; Viviano, R.P.; van Rooden, S.; Ofen, N.; Damoiseaux, J.S. Subjective memory complaints are associated with brain activation supporting successful memory encoding. *Neurobiol. Aging* **2017**, *60*, 71–80. [CrossRef]

22. Hu, X.; Uhle, F.; Fliessbach, K.; Wagner, M.; Han, Y.; Weber, B.; Jessen, F. Reduced future-oriented decision making in individuals with subjective cognitive decline: A functional MRI study. *Alzheimers Dement.* **2017**, *6*, 222–231. [CrossRef] [PubMed]

23. Rodda, J.E.; Dannhauser, T.M.; Cutinha, D.J.; Shergill, S.S.; Walker, Z. Subjective cognitive impairment: Increased prefrontal cortex activation compared to controls during an encoding task. *Int. J. Geriatr. Psychiatry* **2009**, *24*, 865–874. [CrossRef]

24. Rodda, J.; Dannhauser, T.; Cutinha, D.J.; Shergill, S.S.; Walker, Z. Subjective cognitive impairment: Functional MRI during a divided attention task. *Eur. Psychiatry* **2011**, *26*, 457–462. [CrossRef] [PubMed]

25. Dumas, J.A.; Kutz, A.M.; McDonald, B.C.; Naylor, M.R.; Pfaff, A.C.; Saykin, A.J.; Newhouse, P.A. Increased working memory-related brain activity in middle-aged women with cognitive complaints. *Neurobiol. Aging* **2013**, *34*, 1145–1147. [CrossRef] [PubMed]

26. Peters, J.; Buchel, C. Episodic future thinking reduces reward delay discounting through an enhancement of prefrontal-mediotemporal interactions. *Neuron* **2010**, *66*, 138–148. [CrossRef] [PubMed]

27. Lindbergh, C.A.; Puente, A.N.; Gray, J.C.; Mackillop, J.; Miller, L.S. Delay and probability discounting as candidate markers for dementia: An initial investigation. *Arch. Clin. Neuropsychol.* **2014**, *29*, 651–662. [CrossRef] [PubMed]

28. Rabin, L.A.; Smart, C.M.; Crane, P.K.; Amariglio, R.E.; Berman, L.M.; Boada, M.; Buckley, R.F.; Chetelat, G.; Dubois, B.; Ellis, K.A.; et al. Subjective Cognitive Decline in Older Adults: An Overview of Self-Report Measures Used Across 19 International Research Studies. *J. Alzheimers Dis.* **2015**, *48* (Suppl. 1), S63–S86. [CrossRef] [PubMed]

29. Katzman, R. Education and the prevalence of dementia and Alzheimer's disease. *Neurology* **1993**, *43*, 13–20. [CrossRef] [PubMed]

30. Cabeza, R. Hemispheric asymmetry reduction in older adults: The HAROLD model. *Psychol. Aging* **2002**, *17*, 85–100. [CrossRef] [PubMed]

31. Hohwy, J. Attention and conscious perception in the hypothesis testing brain. *Front. Psychol.* **2012**, *3*, 96. [CrossRef] [PubMed]

32. Hohwy, J. Priors in perception: Top-down modulation, Bayesian perceptual learning rate, and prediction error minimization. *Conscious. Cognit.* **2017**, *47*, 75–85. [CrossRef] [PubMed]

33. Rao, R.P.; Ballard, D.H. Predictive coding in the visual cortex: A functional interpretation of some extra-classical receptive-field effects. *Nat. Neurosci.* **1999**, *2*, 79–87. [CrossRef] [PubMed]

34. Botvinick, M.; Nystrom, L.E.; Fissell, K.; Carter, C.S.; Cohen, J.D. Conflict monitoring versus selection-for-action in anterior cingulate cortex. *Nature* **1999**, *402*, 179–181. [CrossRef] [PubMed]

35. Niki, H.; Watanabe, M. Prefrontal and cingulate unit activity during timing behavior in the monkey. *Brain Res.* **1979**, *171*, 213–224. [CrossRef]

36. Botvinick, M.M.; Cohen, J.D.; Carter, C.S. Conflict monitoring and anterior cingulate cortex: An update. *Trends Cognit. Sci.* **2004**, *8*, 539–546. [CrossRef] [PubMed]

37. Friston, K. The free-energy principle: A unified brain theory? *Nat. Rev. Neurosci.* **2010**, *11*, 127–138. [CrossRef] [PubMed]

38. Ebitz, R.B.; Hayden, B.Y. Dorsal anterior cingulate: A Rorschach test for cognitive neuroscience. *Nat. Neurosci.* **2016**, *19*, 1278–1279. [CrossRef] [PubMed]

39. Nelson, T.O. Metamemory: A theoretical framework and new findings. *Psychol. Learn. Motiv.* **1990**, *26*, 125–173.

40. McClelland, J.L.; Rumelhart, D.E. *Explorations in Parallel Distributed Processing: A Handbook of Models, Programs, and Exercises*; MIT Press: Cambridge, MA, USA, 1989.

41. Hebb, D.O. *The Organizations of Behavior: A Neuropsychological Theory*; Lawrence Erlbaum: New York, NY, USA, 1964.

42. Keysers, C.; Gazzola, V. Hebbian learning and predictive mirror neurons for actions, sensations and emotions. *Philos. Trans R. Soc. Lond. B Biol. Sci.* **2014**, *369*, 20130175. [CrossRef]

43. Stern, Y. Cognitive reserve. *Neuropsychologia* **2009**, *47*, 2015–2028. [CrossRef]

44. Stern, Y. Cognitive reserve in ageing and Alzheimer's disease. *Lancet Neurol.* **2012**, *11*, 1006–1012. [CrossRef]

45. McCrae, R.R.; Costa Jr, P.T. Brief versions of the NEO-PI-3. *J. Individ. Differ.* **2007**, *28*, 116–128. [CrossRef]

46. Pailing, P.E.; Segalowitz, S.J. The error-related negativity as a state and trait measure: Motivation, personality, and ERPs in response to errors. *Psychophysiology* **2004**, *41*, 84–95. [CrossRef] [PubMed]

47. Owens, H.; Maxmen, J.S. Mood and affect: A semantic confusion. *Am. J. Psychiatry* **1979**. [CrossRef]

48. Li, Q.; Wang, S.; Milot, E.; Bergeron, P.; Ferrucci, L.; Fried, L.P.; Cohen, A.A. Homeostatic dysregulation proceeds in parallel in multiple physiological systems. *Aging Cell* **2015**, *14*, 1103–1112. [CrossRef] [PubMed]

49. Dodd, P.R.; Scott, H.L.; Westphalen, R.I. Excitotoxic mechanisms in the pathogenesis of dementia. *Neurochem. Int.* **1994**, *25*, 203–219. [CrossRef]

50. Palop, J.J.; Chin, J.; Roberson, E.D.; Wang, J.; Thwin, M.T.; Bien-Ly, N.; Yoo, J.; Ho, K.O.; Yu, G.Q.; Kreitzer, A.; et al. Aberrant excitatory neuronal activity and compensatory remodeling of inhibitory hippocampal circuits in mouse models of Alzheimer's disease. *Neuron* **2007**, *55*, 697–711. [CrossRef]

51. Reisberg, B.; Shulman, M.B.; Torossian, C.; Leng, L.; Zhu, W. Outcome over seven years of healthy adults with and without subjective cognitive impairment. *Alzheimers Dement.* **2010**, *6*, 11–24. [CrossRef]
52. Bhome, R.; Berry, A.J.; Huntley, J.D.; Howard, R.J. Interventions for subjective cognitive decline: Systematic review and meta-analysis. *BMJ Open* **2018**, *8*, e021610. [CrossRef]

Article

Age-Related Deficits in Memory Encoding and Retrieval in Word List Free Recall

Dorina Cadar [1], Marius Usher [2] and Eddy J. Davelaar [3,*

[1] Department of Behavioural Science and Health, University College London, London WC1E 7HB, UK; d.cadar@ucl.ac.uk
[2] Department of Psychology, Tel-Aviv University, Tel-Aviv 69978, Israel; marius@tauex.tau.ac.il
[3] Department of Psychological Sciences, Birkbeck, University of London, London WC1E 7HX, UK
* Correspondence: e.davelaar@bbk.ac.uk; Tel.: +44-207-909-0807

Received: 1 November 2018; Accepted: 29 November 2018; Published: 30 November 2018

Abstract: Although ageing is known to affect memory, the precise nature of its effect on retrieval and encoding processes is not well understood. Here, we examine the effect of ageing on the free recall of word lists, in which the semantic structure of word sequences was manipulated from unrelated words to pairs of associated words with various separations (between pair members) within the sequence. We find that ageing is associated with reduced total recall, especially for sequences with associated words. Furthermore, we find that the degree of semantic clustering (controlled for chance clustering) shows an age effect and that it interacts with the distance between the words within a pair. The results are consistent with the view that age effects in memory are mediated both by retrieval and by encoding processes associated with frontal control and working memory.

Keywords: ageing; memory encoding; memory retrieval; free recall; semantic clustering

1. Introduction

Episodic memory is known to decline with age, and this decline affects some tasks and processes more than others. For example, the largest age-related declines are found in tasks, such as free recall, which depend on retrieval strategies, while smaller deficits are found in recognition memory [1]. This conclusion is supported by studies of free recall [2], which reported that older adults have lower temporal contiguity effects (a reduction in the conditional probability of sequentially reporting items in proximal list positions). This is an effect which is associated with the use of retrieved context to guide subsequent retrieval [3]. Furthermore, an ageing retrieval deficit is suggested by verbal fluency studies, showing that older adults retrieve fewer words in a semantic fluency task, in which as many animal names as possible have to be generated within a fixed time period [4]. As no encoding is required in verbal fluency, this strongly supports an ageing effect on memory retrieval. While such studies suggest that age-related decline in episodic memory is due to a retrieval deficit, encoding deficits have also been demonstrated, as older participants are less likely to form rich, elaborative memory traces [5]. Furthermore, smaller age differences in memory are found when the initial encoding is equated [6].

The ageing deficit has been explained in terms of a reduction in frontal lobe functioning [7]. This hypothesis is supported by significant correlations between neuropsychological measures of frontal lobe functions and memory tests sensitive to ageing [8]. Morphological [9] and neuroimaging studies [10,11] give further support to this hypothesis by showing reduced activation in the left prefrontal cortex (PFC, associated with memory encoding and the processing of semantic information) during memory encoding and semantic processing. Further support for the "ageing as PFC reduction" perspective comes from the observation that like older adults, frontal patients show clear deficits in tests of free recall but not in memory recognition [12,13] and source memory (more than in item memory [14]).

Moscovitch [15] argued that both encoding and retrieval of consciously apprehended information are supported by the medial temporal lobes and hippocampus [16–18], and that these processes are under voluntary control. He noted that the frontal lobes operate on these structures and guide the encoding, retrieval, monitoring, and organisation of information: "By operating on the medial temporal and diencephalic system, the frontal lobes act as working-with-memory structures that control the more reflexive medial temporal and diencephalic system and confer a measure of intelligence and direction to it" ([15], p. 8, see also [19]).

Although evidence suggests that ageing affects both encoding and retrieval, a recurrent problem of interpretation is that different tasks and methods are used to provide support in favour of the encoding or retrieval interpretation. Here, we focus on the task of free recall (a single task from which metrics of encoding and retrieval can be extracted) in which the effect of age can be investigated.

Free recall of word list sequences can be used to quantify the contributions of encoding and retrieval processes in ageing. In a typical experimental paradigm, participants memorise a sequence of words consisting of exemplars from a small number of categories (e.g., four words from each of four categories). The words may be presented sequentially in a blocked-by-category fashion or randomized, and the recall is scored both in terms of the total number of correctly reported words and in terms of the degree of semantic clustering (see Section 2.4). One rationale for examining clustering performance is that it could provide a measure of retrieval strategies and memory organisation [20], uncontaminated by differences in total recall. Ageing studies have shown that semantic clustering scores are lowered for older adults and frontal patients compared with controls [20,21], but the evidence on the ageing effect remains mixed [22,23]. However, differences in semantic clustering scores in free recall may reflect both encoding and retrieval processes [24]. For example, memorising a word is easier when a semantically related word was just committed to memory. In that case, the semantics will help to encode the words in an episodic chunk, which during retrieval leads to a clustered output. In the absence of a semantically supported chunk, having just retrieved a word might prime the retrieval of a semantically related word. This would also produce a clustered output. Both processes are sensitive to ageing effects, but relative sensitivities are unclear.

The aim of our study is to estimate whether age deficits in a single task of word free recall are mediated by retrieval, by encoding processes, or by both. Based on previous studies [21,25], we expect to find age-related differences in both total recall and clustering. These studies, however, were able to assess only encoding or retrieval effects, but not both. To dissociate or isolate encoding and retrieval deficits within a single task, we employed a method that is sensitive to associative processes at encoding, while keeping the retrieval demands constant. This design is created by systematically varying the separation between semantic associates within a word list as part of a free recall paradigm, which is adapted from Glanzer [26] (see also [27]), who presented participants with lists made of pairs of weakly associated words (e.g., "stomach–liver").

The critical manipulation is the separation between associated pair members in the list. There are four conditions: Three related and one unrelated. The related ones are divided into three separation levels, i.e., the number of unrelated intervening words between a pair of associates. Thus, for the related conditions, each list contains one separation condition: separation-0 corresponds to a1, a2, b1, b2...; separation-1 to a1, b1, a2, b2 ... ; and separation-5 to a1, b1, c1, d1, e1, f1, a2, b2 ... . Using such a design, Glanzer [26] reported that memory recall was higher for lists with associated, compared with unrelated words, and was higher when the associates were separated by fewer unrelated words (small separation). As proposed by Glanzer, and demonstrated in simulation models [27,28], this separation effect can be explained as a result of semantic-associative processes during encoding in a capacity-limited working memory [27,29]. We have shown how encoding processes supported by the prefrontal cortex (PFC) operate on sequences that include an associative structure [24]. Specifically, when members of a pair reside in working memory simultaneously, the associative link between the words is increased (see also [30]), by recruiting category units that are linked with the list context and support subsequent retrieval. Thus, total recall, as well as semantic clustering, are predicted to be higher for word pairs whose members are separated by fewer unrelated words, as they are more likely to co-occupy working memory. Such a differential effect observed with a separation design is thus an encoding effect.

Using this paradigm, the separation-1 condition (i.e., one intervening item between pair members) is critical in demonstrating the age deficits that are in part due to declines in working memory assisted encoding. At separation-1, older participants may be more strongly affected in their recall of related words, because those words were less likely to have been maintained within their reduced working memory. Therefore, if ageing affects memory encoding, we predict an age-related reduction in semantic clustering at separation-1, which will differ from separation-5, as recall of word pairs separated by five intervening words is not assumed to be mediated by working memory. Thus, with this paradigm, we can demonstrate the additional influence of encoding processes on top of retrieval within a single task, based on the separation between associates.

To summarise, we expect age differences at encoding to result in an interaction between age and separation, either on total recall or on clustering, while age differences at retrieval to result in an interaction between age and relatedness with a long-separation (separation-5).

2. Materials and Methods

2.1. Participants

Forty native English speakers, who reported being in good health, took part in this study. The younger (n = 20, age range 19–35, mean 27) and older (n = 20, age range 55–65, mean 61) groups were balanced in educational background (young: 55% high-school; 45% university; older: 50% high-school, 50% university). Both groups were tested on a Test Of English as Foreign Language (TOEFL) as part of a standard test of general knowledge. The Quick test format used in the study was composed of 10 words, each followed by a four-alternative multiple choice question probing the word's meaning. The vocabulary test was administered after the free recall memory task. The mean score for young adults was 6.05 (SD = 2.19) and for the older adults it was 5.4 (SD = 2.52). No group differences were present (t < 1). Thus, any group differences in memory recall or clustering cannot be attributed to group differences in background vocabulary. The study was approved by the Birkbeck ethics committee.

2.2. Materials

For the memory task, we used a pool of 240 words consisting of common one- and two-syllable nouns with 90 pairs of weak associates (e.g., "stomach–liver", see [28]); weak associates were used in order to avoid guessing strategies. The words have a mean written word frequency of 66.9 per million, and the average associative strength was 0.15 (SD = 0.14; [31]). Each memory list consisted of 12 words and was constructed in accordance with one of four separation conditions. The five lists in the unrelated condition contained twelve words that did not have a semantic relation. We created three types of related lists by varying the number of intervening unrelated n items separating the members of a pair. We used separation-0, -1, and -5. With separation-0, the members of an associated pair, a1–a2, appeared in temporally adjacent list positions (e.g., a1, a2, b1, b2, c1, c2, d1, d2, e1, e2, f1, f2). In separation-1, the pair members were separated by one unrelated word, which was a member of a different pair (e.g., a1, b1, a2, b2, c1, d1, c2, d2, e1, f1, e2, f2), whereas in the separation-5 condition, five words separated the pair members (e.g., a1, b1, c1, d1, e1, f1, a2, b2, c2, d2, e2, f2). There were a total of 23 lists: three practice trials and 20 experimental trials.

2.3. Procedure

Participants were tested individually in noise-attenuated conditions and presented with the computerised memory task followed by the vocabulary test. Instructions were presented on the computer screen as well as verbally to ensure comprehension. Presentation of the memory test was visual. Participants were seated in front of the monitor, fixating the centre of the screen. Each trial started with a row of three question marks in the middle of the screen. These were replaced by words presented at a rate of one word per second. Participants were required to read each word silently and immediately after the final word to perform the arithmetic distractor task. This distractor task consisted of a mix of 12 additions and subtractions of the form A +/− B = C, where A, B and C were positive single digit numbers such as e.g., "1 + 2 = 3". The participant was instructed to quickly press the "k" (when correct) or "s" button on the keyboard to indicate the accuracy of the mathematical expression. A row of three question marks prompted the participants to recall aloud as many words from the list as possible, in any order. The experimenter wrote down the recalled items.

2.4. Semantic Clustering Scores

Semantic clustering was estimated using the pair frequency score, which is identical to the original clustering metric proposed by Bousfield and Bousfield [32]. This metric looks at the memory recall protocol and focuses on the number of within-category repetitions. To illustrate, consider a two-category word list, made of *cat, dog, rabbit, fork, knife, spoon*, with a recall output of *cat, dog, fork, knife, rabbit*. The pair frequency is the difference between the observed number of within-cluster transitions, which is 2 (cat → dog, fork → knife) and the expected number of within-cluster transitions, which is calculated using the following formula:

$$\sum_{i=1}^{N_c} \frac{n_i(n_i - 1)}{r} \tag{1}$$

where n_i is the number of correct words for category i, out of N_c recalled categories, and r is the total recall. In the example, $i = 2$, $n_1 = 3$ (cat, dog, rabbit), $n_2 = 2$ (fork, knife), and r is 5. This leads to an expected number of within-cluster transitions of $3 \times 2/5 + 2 \times 1/5 = 1.6$, and the overall clustering score of $2 - 1.6 = 0.4$.

In addition, we also estimated semantic clustering, using the California Verbal Learning Test (CVLT: [33]) measure. As the results are the same, we only report the pair-frequency measure here.

3. Results

Figure 1A,B show the average number of words recalled in the four conditions (three separations and one unrelated) and for the related (averaged across separation) and unrelated conditions, for young and older adults, respectively. Participants recalled more words in the related than in the unrelated condition. Younger adults reported more words than older adults, especially in the related condition. A 2 (age: young, older) × 2 (association: related, unrelated) mixed ANOVA revealed a main effect of age (F (1,38) = 15.83, MSe = 1.16, $p < 0.001$, $\eta^2 = 0.29$), a main effect of association (F (1,38) = 102.85, MSe = 0.31, $p < 0.001$, $\eta^2 = 0.73$), and an interaction (F (1,38) = 5.23, MSe = 0.31, $p < 0.05$, $\eta^2 = 0.12$). The interaction was due to a larger increase in total recall with related lists for younger than for older adults (t (38) = 2.29, $p < 0.05$). The same conclusions were obtained when the ANOVA was limited to unrelated versus separation-5.

Both groups showed equivalent decrease in recall performance with increasing separation between pair members. A 2 (age: young, older) × 3 (separation: 0, 1, 5) mixed ANOVA revealed a significant main effect of age (F (1,38) = 18.42, MSe = 2.52, $p < 0.001$, $\eta^2 = 0.33$) and a main effect of separation (F (2,76) = 12.25, MSe = 0.74, $p < 0.001$, $\eta^2 = 0.24$). The interaction between age and separation was not significant (F < 1).

We computed, for each separation, a semantic clustering score. Several metrics have been developed to measure semantic clustering [20,34]. Here, we use the pair frequency, which is identical to the original clustering metric proposed by Bousfield and Bousfield [32]. If, for example, a participant saw the words light$_1$ man$_2$ candle$_1$ father$_2$ hat$_3$ river$_4$ cap$_3$ lake$_4$ door$_5$ thief$_6$ wall$_5$ crook$_6$ (separation-1) and then reported lake$_4$ river$_4$ light$_1$ man$_2$ candle$_1$ (numbers are only added for illustration and were not present in the experiment), the clustering score would be 0.20 (OBS − EXP = 1 − (2 × 1/5 + 2 × 1/5 + 1 × 0/5) = 0.2). See Methods for details.

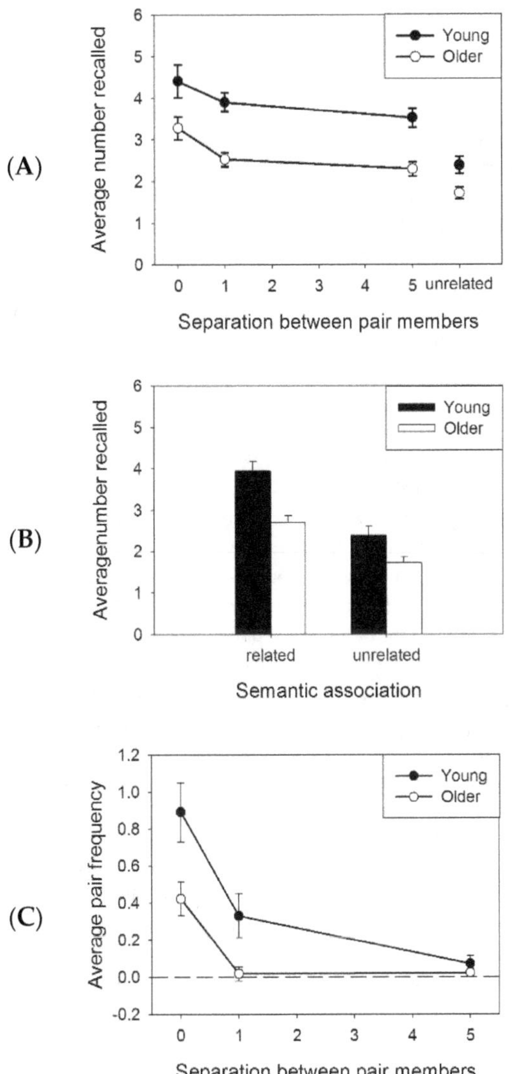

Figure 1. (A) Average number of words recalled as a function of the separation between the members of an associated pair and age. The separations in the related condition are presented. (B) The same as in panel A, but the related condition is averaged across separations. (C) Clustering scores in the related conditions as a function of the separation between pair members and age. The error bars represent standard error of the mean.

Figure 1C presents the clustering scores for each separation. Younger adults (filled circles) clustered the members of a pair more often than older adults (open circles). In addition, increasing the separation between pair members during encoding lowered the clustering effect for both age groups. A 2 (age: younger, older) × 3 (separation: 0, 1, 5) mixed ANOVA revealed a significant main effect of separation (F (2,76) = 39.39, MSe = 0.11, $p < 0.001$, $\eta^2 = 0.51$) and age (F (2,76) = 7.34, MSe = 0.31, $p < 0.01$, $\eta^2 = 0.16$) and an interaction between separation and age (F (2,76) = 4.29, MSe = 0.11, $p < 0.05$, $\eta^2 = 0.10$). This interaction was due to an absence of an age effect for separation 5 ($p > 0.34$). Also, older adults did not cluster the words when the pair members were separated by one or more unrelated words (all $ps > 0.10$).

4. Discussion

We tested for age-related differences in memory recall of word lists made of associated pairs, with various degrees of separation (number of unrelated intervening words) between the pair members. First, we found that lists containing associated pairs are recalled better than lists of un-associated words in both age groups and that this difference is larger for young adults. In particular, we found only small age differences for lists of unrelated words (0.67 words) and we found larger age differences for lists containing associated pairs (1.32 words). This interaction is also present when comparing the unrelated and separation-5 conditions. As word associates, which are separated by five intervening unrelated words, are not likely to co-occupy working memory, this interaction is likely to indicate a retrieval deficit: younger participants might be better in using the last word recalled, as a cue to memory search [2].

Second, we found that for both age groups, the association effect (better recall of associated than unrelated words) also depends on the separation of the pair members in the memory list, with better performance at short separation. This replicates the results by Glanzer [26], indicating that the encoding of the list is more effective when associated pairs co-occupy a capacity-limited working memory [27,29]. If older participants have reduced working memory capacity, one could expect that they will benefit less of associated pairs at separation-1; in both conditions, the pairs are unlikely to co-occupy working memory. Although the difference in total recall for separations 1 and 5 was numerically larger for young than for older participants, this difference was not statistically significant.

Third, we examined the clustering measures, of the two age groups, at each level of separation. Participants produced more clustered output, at short separations, consistent with an encoding mechanism (co-occupation of the associates in working memory, enhances their association). As we predicted, younger participants produced more clusters at separation-1 than at separation-5, consistent with an age-dependent working memory-assisted encoding. The clustering frequency for the old group at separation-1 was not different from that at separation-5, which was at chance level. Thus, older adults show reduced beneficial effect of closer temporal proximity, as expected from a reduced encoding of relations between words that co-occupy working memory. The absence of clustering at separation-1 critically supports the suggestion that older adults' working memory capacity is just too small to have both members of a pair active simultaneously after a single intervening unrelated item. This prevents the detection of the semantic similarity and the associative boost that is needed for these items to be recalled consecutively. While we report the pair frequency measure for semantic clustering, we also looked at other clustering measures such as the one used for the CVLT. Age-effects at separation-1 were present in these measures at separation-1, with older adults showing no clustering beyond what is expected by chance.

We suggest that the results indicate age effects in both the retrieval and encoding of a list of words, which are due to a reduced ability to encode relations between list-words (see also [35] for similar findings using a memory recognition paradigm). This is consistent with previous age studies of clustering, which used the CVLT [33], reporting marked age-related decline [25]. In a previous study, we have shown that ageing is associated with a significant deficit in a test that shares some features with the CVLT—The conceptual-span task [36]. Furthermore, our results parallel those obtained in memory studies in frontal patients, who do not show enhanced recall with related, compared with unrelated word lists [37], and who do not spontaneously utilise cues to enhance their memory encoding and retrieval, and show reduced semantic clustering [12]. These frontal memory deficits are well accounted for by neuropsychological models of memory, such as Hemispheric Encoding-Retrieval Asymmetry

(HERA) [38], which postulate a role for the left PFC in semantic processing and memory encoding and by computational models that unpack the PFC mechanisms involved in memory encoding and retrieval [24,39]. For example, in the Categorization–Activation–Novelty (CAN; [24]) model, the PFC detects the semantic differences and similarities among list items that co-occupy WM. Semantic associates get an encoding boost, but only if their relatedness has been detected.

We believe that these similarities are explained best by the hypothesis that ageing deficits in memory are caused by a frontal mechanism [7,40]. As previously reviewed, this hypothesis is supported by neuropsychological, neurological, and neuroimaging evidence, showing changes in frontal lobe structures and functioning with advancing age. This interpretation is also consistent with the finding that older adults are less likely to form rich, elaborative memory traces [5]. The results presented here provide convergent evidence within a single paradigm suggesting that age-related declines in total recall are associated with a decrease in the use of associative information during both encoding and retrieval.

Author Contributions: Formal analysis, D.C., M.U., and E.J.D.; Investigation, D.C.; Methodology, D.C., M.U., E.J.D.; Writing—Original draft, D.C., M.U., and E.J.D.

Funding: D.C. is currently funded by the National Institute on Aging (Grant RO1AG7644) and the Economic and Social Research Council (ESRC). M.U. is supported by the Israeli Science Foundation (1413/17) and the Binational (Israel-USA) Science Foundation (2014612).

Conflicts of Interest: The authors declare no conflict of interest.

References

1. Craik, F.I.M.; McDowd, J.M. Age differences in recall and recognition. *J. Exp. Psychol. Learn. Mem. Cogn.* **1987**, *13*, 474–479. [CrossRef]
2. Kahana, M.J.; Howard, M.W.; Zaromb, F.; Wingfield, A. Age dissociates recency and lag recency effects in free recall. *J. Exp. Psychol. Learn. Mem. Cogn.* **2002**, *3*, 530–540. [CrossRef]
3. Howard, M.W.; Kahana, M.J. A distributed representation of temporal context. *J. Math. Psychol.* **2002**, *46*, 269–299. [CrossRef]
4. Wingfield, A.; Lindfield, K.C.; Kahana, M.J. Adult age differences in the temporal characteristics of category free recall. *Psychol. Aging* **1998**, *13*, 256–266. [CrossRef]
5. Craik, F.I.M.; Jennings, J.M. Human memory. In *Handbook of Aging and Cognition*, 1st ed.; Craik, F.I.M., Salthouse, T.A., Eds.; Erlbaum: Hillsdale, NJ, USA, 1992; pp. 51–110.
6. Rybarczyk, B.D.; Hart, R.P.; Harkins, S.W. Age and forgetting rate with pictorial stimuli. *Psychol. Aging* **1987**, *2*, 404–406. [CrossRef] [PubMed]
7. West, R. An application of prefrontal cortex function theory to cognitive aging. *Psychol. Bull.* **1996**, *120*, 272–292. [CrossRef] [PubMed]
8. Parkin, A.J.; Walter, B.M. Recollective experience, normal aging, and frontal dysfunction. *Psychol. Aging* **1992**, *7*, 290–298. [CrossRef] [PubMed]
9. Haug, H.; Eggers, R. Morphometry of the human cortex cerebri and human striatum during aging. *Neurobiol. Aging* **1991**, *12*, 336–338. [CrossRef]
10. Cabeza, R.; Grady, C.L.; Nyberg, L.; McIntosh, A.R.; Tulving, E.; Kapur, S.; Jennings, J.M.; Houle, S.; Craik, F.I.M. Age-related differences in neural activity during memory encoding and retrieval: A positron emission tomography study. *J. Neurosci.* **1997**, *17*, 391–400. [CrossRef]
11. Stebbins, G.T.; Carrillo, M.C.; Dorman, J.; Dirksen, C.; Desond, J.E.; Turner, D.A.; Bennett, D.A.; Wilson, R.S.; Glover, G.; Gabrieli, J.D. Aging effects on memory encoding in the frontal lobes. *Psychol. Aging* **2002**, *17*, 44–55. [CrossRef]
12. Baldo, J.V.; Delis, D.; Kramer, J.; Shimamura, A.P. Memory performance on the California Verbal Learning Test II: Finding from patients with local frontal lesions. *J. Int. Neuropsychol. Soc.* **2002**, *8*, 539–546. [CrossRef] [PubMed]
13. Janowsky, J.S.; Shimamura, A.P.; Squire, L.R. Source memory impairment in patients with frontal lobe lesions. *Neuropsychologia* **1989**, *27*, 1043–1056. [CrossRef]

14. Simons, J.S.; Verfaellie, M.; Galton, C.J.; Miller, B.L.; Hodges, J.R.; Graham, K.S. Recollection-based memory in frontotemporal dementia: Implications for theories of long-term memory. *Brain* **2002**, *125*, 2523–2536. [CrossRef]

15. Moscovitch, M. Amnesia. In *The International Encyclopedia of Social and Behavioural Sciences*; Smesler, N.B., Baltes, O.B., Eds.; Pergamon/Elsevier Science: Oxford, UK, 2004; pp. 1–26.

16. Moscovitch, M.; Westmacott, R.; Gilboa, A.; Addis, D.R.; Rosenbaum, R.S.; Viskontas, I.; Priselac, S.; Svoboda, E.; Ziegler, M.; Black, S.; et al. Hippocampal complex contribution to retention and retrieval of recent and remote episodic and semantic memories: Evidence from behavioral and neuroimaging studies of healthy and brain-damaged people. In *Dynamic Cognitive Processes*; Ohta, N., MacLeod, C.M., Uttl, B., Eds.; Springer: Tokyo, Japan, 2005; pp. 333–380.

17. Nadel, L.; Moscovitch, M. Memory consolidation, retrograde amnesia and the hippocampal complex. *Curr. Opin. Neurobiol.* **1997**, *7*, 217–227. [CrossRef]

18. Nadel, L.; Samsonovich, A.; Ryan, L.; Moscovitch, M. Multiple trace theory of human memory: Computational, neuroimaging, and neuropsychological results. *Hippocampus* **2000**, *10*, 352–368. [CrossRef]

19. Simons, J.S.; Spiers, H.J. Prefrontal and medial temporal lobe interactions in long-term memory. *Nat. Rev. Neurosci.* **2003**, *4*, 637–648. [CrossRef] [PubMed]

20. Stricker, J.L.; Brown, G.G.; Wixted, J.T.; Baldo, J.V.; Delis, D. New semantic and serial clustering indices for the California Verbal Learning Test 2: Background, rationale, and formulae. *J. Int. Neuropsychol. Soc.* **2002**, *8*, 425–435. [CrossRef] [PubMed]

21. Taconnat, L.; Raz, N.; Tocze, C.; Bouazzaoui, B.; Sauzeon, H.; Fay, S.; Isingrini, M. Ageing and organisation strategies in free recall: the role of cognitive flexibility. *Eur. J. Cogn. Psychol.* **2009**, *21*, 347–365. [CrossRef]

22. Backman, L.; Larsson, M. Recall of organizable words and objects in adulthood: Influences of instructions, retention interval, and retrieval cues. *J. Gerontol.* **1992**, *47*, P273–P278. [CrossRef] [PubMed]

23. Verhaeghen, P.; Marcoen, A. More or less the same? A memorability analysis on episodic memory tasks in young and older adults. *J. Gerontol.* **1993**, *48*, P172–P178. [CrossRef] [PubMed]

24. Elhalal, A.; Davelaar, E.J.; Usher, M. The role of the frontal cortex in memory: An investigation of the Von Restorff effect. *Front. Hum. Neurosci.* **2014**, *8*, 410. [CrossRef] [PubMed]

25. Norman, M.A.; Evans, J.D.; Miller, S.W.; Heaton, R.K. Demographically corrected norms for the California Verbal Learning Test. *J. Clin. Exp. Neuropsychol.* **2000**, *22*, 80–94. [CrossRef]

26. Glanzer, M. Distance between related words in free recall: Trace of the STS. *J. Verbal Learn. Verbal Behav.* **1969**, *8*, 105–111. [CrossRef]

27. Davelaar, E.J.; Haarmann, H.J.; Goshen-Gottstein, Y.; Usher, M. Semantic similarity dissociates short from long-term recency effects: Testing a neurocomputational model of list memory. *Mem. Cogn.* **2006**, *34*, 323–334. [CrossRef]

28. Haarmann, H.J.; Usher, M. Maintenance of semantic information in capacity-limited item short-term memory. *Psychol. Bull. Rev.* **2001**, *8*, 568–578. [CrossRef]

29. Cowan, N. The magical number 4 in short-term memory: A reconsideration of mental storage capacity. *Behav. Brain Sci.* **2001**, *24*, 87–185. [CrossRef]

30. Raaijmakers, J.G.W.; Shiffrin, R.M. Search of associative memory. *Psychol. Rev.* **1981**, *88*, 93–134. [CrossRef]

31. Nelson, D.L.; McEvoy, C.L.; Schreiber, T.A. The University of South Florida word association, rhyme, and word fragment norms. Available online: http://www.usf.edu/FreeAssociation/ (accessed on 30 January 2008).

32. Bousfield, A.K.; Bousfield, W.A. Measurement of clustering and of sequential constancies in repeated free recall. *Psychol. Rep.* **1966**, *19*, 935–942. [CrossRef]

33. Delis, D.C.; Kramer, J.H.; Kaplan, E.; Ober, B. *California Verbal Learning Test: Adult Version Manual*, 1st ed.; The Psychological Corporation: San Antonio, TX, USA, 1987.

34. Sternberg, R.J.; Tulving, E. The measurement of subjective organization in free recall. *Psychol. Bull.* **1977**, *84*, 539–556. [CrossRef]

35. Naveh-Benjamin, M. Adult age differences in memory performance: Tests of an associative deficit hypothesis. *J. Exp. Psychol. Learn. Mem. Cogn.* **2000**, *26*, 1170–1187. [CrossRef]

36. Haarmann, H.J.; Ashling, G.E.; Davelaar, E.J.; Usher, M. Age-related declines in context maintenance and semantic short-term memory. *Q. J. Exp. Psychol.* **2005**, *58*, 34–53. [CrossRef] [PubMed]

37. Hirst, W.; Volpe, B.T. Memory strategies with brain damage. *Brain Cogn.* **1988**, *8*, 379–408. [CrossRef]

38. Tulving, E.; Kapur, S.; Craik, F.I.M.; Moscovitch, M.; Houle, S. Hemispheric encoding/retrieval asymmetry in episodic memory: Positron emission tomography findings. *Proc. Natl. Acad. Sci.* **1994**, *91*, 2016–2020. [CrossRef] [PubMed]
39. Becker, S.; Lim, J. A computational model of prefrontal control in free recall: Strategic memory use in the California Verbal Learning Task. *J. Cogn. Neurosci.* **2003**, *15*, 821–832. [CrossRef]
40. Moscovitch, M.; Winocur, G. Frontal lobes, memory and aging. *Ann. N. Y. Acad. Sci.* **1995**, *769*, 119–150. [CrossRef]

Article

Active Navigation in Virtual Environments Benefits Spatial Memory in Older Adults

Melissa E. Meade [1],*, John G. Meade [1], Hélène Sauzeon [2,3] and Myra A. Fernandes [1]

[1] Department of Psychology, University of Waterloo, Waterloo, ON N2L 3G1, Canada; jon.meadee@gmail.com (J.G.M.); mafernan@uwaterloo.ca (M.A.F.)

[2] Lab EA413-Handicap, Activité, Cognition & Santé, Université de Bordeaux, F-33000 Bordeaux, France; helene.sauzeon@u-bordeaux.fr

[3] Flowers Team-INRIA Center, F-33405 Talence, France

* Correspondence: mmeade@uwaterloo.ca; Tel.: +1-519-888-4567 (ext. 37776)

Received: 30 January 2019; Accepted: 21 February 2019; Published: 26 February 2019

Abstract: We investigated age differences in memory for spatial routes that were either actively or passively encoded. A series of virtual environments were created and presented to 20 younger (Mean age = 19.71) and 20 older (Mean age = 74.55) adults, through a cardboard viewer. During encoding, participants explored routes presented within city, park, and mall virtual environments, and were later asked to re-trace their travelled routes. Critically, participants encoded half the virtual environments by passively viewing a guided tour along a pre-selected route, and half through active exploration with volitional control of their movements by using a button press on the viewer. During retrieval, participants were placed in the same starting location and asked to retrace the previously traveled route. We calculated the percentage overlap in the paths travelled at encoding and retrieval, as an indicator of spatial memory accuracy, and examined various measures indexing individual differences in their cognitive approach and visuo-spatial processing abilities. Results showed that active navigation, compared to passive viewing during encoding, resulted in a higher accuracy in spatial memory, with the magnitude of this memory enhancement being significantly larger in older than in younger adults. Regression analyses showed that age and score on the Hooper Visual Organizational test predicted spatial memory accuracy, following the passive and active encoding of routes. The model predicting accuracy following active encoding additionally included the distance of stops from an intersection as a significant predictor, illuminating a cognitive approach that specifically contributes to memory benefits in following active navigation. Results suggest that age-related deficits in spatial memory can be reduced by active encoding.

Keywords: aging; spatial memory; active exploration; virtual reality

---

1. Active Navigation in Virtual Environments Benefits Older Adults' Spatial Memory

Spatial navigation deficits are frequently experienced in the older adult population [1–3]. For example, age-related deficits are commonly observed in the memory for locations, landmarks, routes, and maps [2,4]. Such declines in navigation and spatial memory are critical, because they pose a potential threat to independent living for older adults [5–9], prompting the need to find ways to lessen these age-related deficits. The assessment of age-related impairments in spatial cognition has been facilitated by recent advancements in virtual reality (VR) technology [10], allowing for the efficient examination of navigation performance in virtual environments. In the current study, we took advantage of VR methodology to investigate ways for improving spatial memory performance in a navigation task, in older and younger adults. Specifically, the goal of the current study was to determine whether age-related differences in memory for virtual routes are reduced following active exploration, compared to passive guidance. Additionally, we examined whether individual differences in visuo-spatial ability, and cognitive indices during route exploration, meaningfully contributed to spatial memory performance.

2. Active vs. Passive Encoding of Spatial Information

The term active navigation refers to a condition wherein the participant has motor and/or volitional control of their movement through an environment, whereas passive navigation generally consists of a guided tour [11]. Active navigation has been described to contain both cognitive (mental manipulation of spatial information, allocation of attention, and decision-making) and physical (motor control for locomotion and proprioceptive and vestibular sensory information) components [11], which combine to create rich multi-modal associations that benefit subsequent memory performance [12]. A variety of past studies in young adults suggest significant spatial memorial benefits when encoding is active, rather than passive [13–16].

The memory benefit from active, relative to passive, navigation can be conceptualized as a subject-performed task effect (SPT effect) [17,18]. The SPT effect describes a pattern of superior memory performance when encoding involves the direct engagement of the participant, such as when performing rather than watching an action that is associated with a to-be-remembered word [19]. Importantly, the SPT effect remains unchanged with aging, enhancing memory performance in older adults [20–22]. The encoding tasks that involve an SPT have been described to have a high degree of environmental support, accounting for a boost in memory in seniors [23,24]. With regard to encoding manipulations, the environmental support refers to a situation in which older adults do not need to self-initiate encoding operations. By this account, active navigation should provide a good form of environmental support, as it involves the SPT of physically and cognitively engaging with the to-be-remembered spatial information. Past studies, however, have reported inconsistent results regarding the benefits of active relative to passive navigation in older adults. Specifically, benefits from active navigation are sometimes observed in older adults [15,25,26], but there is also evidence showing that active encoding hinders memory performance in seniors [2,27].

The degree of involvement of both decision-making processes and motor processing in active exploration conditions could be critical for explaining the inconsistencies regarding age differences in active/passive manipulation. It is well-known that older adults have to allocate additional cognitive resources to control their locomotion [28,29], suggesting that the physical components of active navigation are unlikely to be the reason for any memory benefit in seniors. Indeed, active navigation with low levels of motor control benefits spatial memory, whereas a high degree of motoric requirement creates an additional cognitive load that impairs memory in older adults [27].

Some have suggested that the cognitive components of active navigation benefit the memory (for directions and the locations of landmarks), more so than physical motor processing [29–31]. Decision-making in particular has been shown to be beneficial in older adults for episodic memory of landmarks and objects [15] and it may elicit a self-reference effect that is known to benefit older adults' spatial memory [32–34]. In other words, the engagement of cognitive and decision-making processes during active navigation conditions could be particularly beneficial for older adults. Likewise, factors related to decision-making, such as intrinsic motivation or curiosity [35,36] may also contribute to the memory benefit from active exploration. As recently claimed by [37], momentary feelings of curiosity can help older adults' memories and executive functions, because the phasic activation of the noradrenergic and dopaminergic systems modulate hippocampal activity [38], facilitating memory and learning. As such, it may be that the cognitive components of active navigation drive the observed active encoding memory benefits in older adults.

Related to this, it is possible that the cognitive processes engaged by the active exploration of an environment allow for more unique associations to be formed with landmarks, and thus enhance the memory for the links between certain environments and paths travelled. Others have suggested that such relational knowledge is critical, as retrieval becomes more likely when the target is part of a rich network of associations, as opposed to being encoding in isolation [12]. This relational encoding can take the form of distinctive sensorimotor associations, and increased elaboration during encoding, due to goal-directed exploration. By actively encoding an experience, and thereby increasing the amount of associations that are credited to an episodic memory, later memory recall can be expected to be improved [12].

3. Using Virtual Reality to Examine Spatial Memory

In recent years, VR technology has become widely used for studying spatial cognition and navigation performance in both younger [39–42], and older adults [10,12,43]. One of the major advantages of VR is the high degree of experimental control that is afforded, to investigate the cognitive and behavioral components that are involved in spatial navigation. For example, VR provides the ability to control for variables that differ between age groups, such as physical fatigue and mobility issues during navigation. Furthermore, the use of VR in the current study allowed for the creation of a series of environments with controlled parameters, including environment size, number of intersections, and visual content, as well as the novel approach of examining spatial memory following trials of either active or passive encoding. Additionally, we developed a means to precisely measure the overlaps in the routes travelled at encoding and retrieval, to provide a measure of memory accuracy in re-tracing a previously travelled route.

Research into the effects of aging on route navigation have largely focused on differences in learning performance and the assessment of knowledge for a highly familiar environment. However, it is not the case that we will only ever need to navigate through familiar environments; rather we frequently encounter new spaces by we need to successfully navigate after brief encounters. As such, in the present study, we took the novel approach of investigating the age difference in memory for routes that were previously travelled on only one occasion. We created a set of virtual environments in which participants traveled routes, and then later re-traced those routes to assess spatial memory performance. During the encoding, participants either traveled routes by actively exploring through volitional control of their own movements, by passively viewing a guided tour. This paradigm was analogous to drivers and passengers in a car, with active trials affording the control that a driver has in deciding on the route and manipulating movement, whereas the passive trials are akin to a passenger simply viewing the route travelled within the environment. We expected that our active encoding manipulation would provide a higher degree of environmental support, as it constitutes a SPT; given that environmental support is particularly helpful to seniors [23,24], active versus passive encoding should benefit older adults' memories more so than younger adults.

4. Method

4.1. Participants

Twenty-two younger (M_{age} = 19.71, SD = 2.19, 11 females) and 22 older (M_{age} = 74.55, SD = 7.82, 10 females) adults participated in our study. Younger adults were undergraduate students at the University of Waterloo who signed up for the study and received partial course credit for their participation. Older adults were recruited from the Waterloo Research in Aging Participant (WRAP) database, and received $10.00 remuneration for their participation. WRAP is a database of healthy seniors residing in the Kitchener–Waterloo area recruited by means of newspaper ads, flyers, and local television segments to participate in research studies taking place on campus at the University of Waterloo. All older adults completed the Montreal Cognitive Assessment (MoCA) to assess cognitive decline [44]. The average MoCA score was 26.95 out of 30 (SD = 2.29), with a range of 20 to 30. Two participants fell below the cut-off a score of 26 (obtaining scores of 24 and 20); which was generally used to indicate normal healthy aging; however, we opted to include all older adults who were tested, as we are not aiming to make a distinction between normal healthy aging and MCI in the current study.

4.2. Materials

VR equipment. The VR program was run on an Asus© Zenfone 3 Laser ™ (see specificities on www.asus.com, Asus, Taipei, Taiwan) and a lightweight hand-held cardboard VR headset for viewing. The VR application run on the Android phone was designed using the 3D engine Unity© (ver. 2017.2.0.f3, Unity Technologies ApS, San Francisco, CA, USA). Twelve virtual environments (VE) were built according to three styles of: city streets, mall hallways, and park trails (see Figure 1 for samples). The virtual environments were topographically similar, with an average area of 234 square meters. Additionally, each environment had six intersections, which is a point where 3+ roads converge.

Figure 1. First-person view and bird's-eye view of the three virtual environment styles. For each style, we created 4 exemplars, though only one from each is shown here. The Stars are visible in the environments during encoding, but not during retrieval. Participants only experienced the environments from a first-person perspective.

Neuropsychological assessments. Spatial visualization ability was evaluated with the Hooper Visual Orientation test [45], which had good internal reliability, as indicated by a coefficient α of 0.88 [46]. This test is scored out of 30; it contains 30 line drawings of common objects that are portrayed as having been cut up and misaligned; participants had to mentally rotate and piece together the visual information, to identify the depicted object. Performance on the Hooper test was included to determine whether spatial visualization abilities were related to, and predictive of, spatial memory performance following active and passive encoding.

The spatial orientation ability was measured with the Santa Barbara Sense-of-Direction scale, which has good internal reliability, as indicated by a coefficient α of 0.88 [47]. For this, the test participants rated their endorsement of 15 statements about spatial orientation abilities, such as "I am good at giving directions', indicating whether they strongly agreed or disagreed, using a 7-point scale. After reverse scoring the necessary items, the responses were summed and divided by 15 to provide an average score of between 1–7 [47]. This measure was included to determine whether the perceived spatial orientation abilities were related to spatial memory performance following active and passive encoding.

5. Procedure

The study received clearance from the ethics review board at the University of Waterloo; all participants gave their written informed consent prior to beginning the experiment. Throughout the entire experiment, participants were seated in comfortable chairs that could turn a full 360° circle, which was placed in the center of a testing room. Following a verbal explanation of how to use the VR headset, participants were given unlimited time to complete a training phase with the VR equipment designed to help them learn how to move in the virtual space. Specifically, they were told that to move forward in VR, they should press a button on the headset, and to change their direction within VR, they were to physically rotate in the spinning chair with their legs. Participants were advised to take as much time as needed in the training phase, to become comfortable with the environment and the movement mechanism. Immediately following the training phase, the experimental session began.

In the experimental session, participants completed 12 cycles of encoding-filler-retrieval phases. The encoding phase was 30 seconds in duration. Once 30 seconds elapsed, the screen turned black and the participant immediately performed a filler task which involved counting backwards by 7's from 100, for 20 seconds, in order to reduce recency effects [48]. In the retrieval phase, participants were placed within the VR environment at the same starting point as the environment they had just experienced. Participants were told that they had 30 seconds to re-trace the exact path just previously travelled in the environment. Participants were given a 1- to 5-min break halfway through the experimental session (i.e., after completing six encoding-filler-retrieval cycles), resuming when they indicated they were ready.

Active vs. passive manipulation. The within-subjects active/passive navigation manipulation was implemented during the encoding phase. Of the 12 virtual environments, six were actively encoded, and six were passively encoded. The order of active and passive trials was mixed, with an instruction appearing at the beginning of each new encoding-filler-retrieval cycle, indicating whether the participant was to explore the environment (active) or watch a guided tour (passive). Participants knew that they would subsequently be asked to retrace the travelled route. In the active trials, the participant had full volitional control of their movement along the paths in the environment, and they were free to choose the route they traveled. To motivate participants to continue to move and explore the environment in the active condition (and not to simply walk back and forth so that their route was later easier to remember) they were told to 'collect' as many star icons along the path as possible, by walking over the stars. Each environment had 10 stars dispersed throughout the environment, along the paths (as shown in Figure 1). On the passive trials, the participant did not control their movement, or route, in the environment, and instead experienced a recording from the active exploration of a previous participant. Using a previous participant's active exploration for the passive trials, in this yoked control manner, was purposely done to equate route length, pauses along the path, and the number of turns, across the active and passive trials.

6. Results

Participant Characteristics. While age differences in spatial abilities were of primary interest in the current study, we also wanted to determine whether any sex differences that we presented in our data, given the previous work, suggested that males perform better than females on a variety of spatial processing measures [49]. To determine whether there were differences between the age groups and sexes, on the neuropsychological assessments included in the study, we ran two 2 (Age: younger, older) × 2 (Sex: male, female) ANOVAS separately, for the performance on the Hooper Visual Orientation test [45] and on the Santa Barbara Sense-of-Direction scale [47]. The analysis on the Hooper Visual Organization performance revealed a main effect of Age, $F(1, 40) = 26.12$, $MSE = 0.26$, $p < 0.001$, and $\eta^2 = 0.40$, such that the scores were higher for younger than for older adults, but there was no effect of Sex, $F(1, 40) = 2.09$, $MSE = 0.02$, $p = 0.16$, $\eta^2 = 0.05$. There were no main effects of Age, $F(1, 40) = 0.06$, $MSE = 0.14$, $p = 0.80$, $\eta^2 = 0.002$, or Sex, $F(1, 40) = 0.48$, $MSE = 1.07$, $p = 0.50$, $\eta^2 = 0.01$ on the Santa Barbara Sense-of-Direction scale, and no interactions on either measure.

Calculation of Route Overlap. The dependent variable of interest was the accuracy in re-tracing the same route at retrieval that had been traveled during encoding. To obtain this measure of spatial memory performance, the overlap between the routes travelled by each participant at encoding and retrieval, for each trial type, was calculated, to produce a 'percent overlap' value for each environment. The percent overlap values were computed by using the Python programming language (https://www.python.org/), Python Software Foundation, Wilmington, DE, USA), to first normalize the routes travelled (during both encoding and retrieval) against the established paths in the environments, and then to compare those normalized routes. The process of normalizing the participant's paths involved removing positional data points that corresponded to the subject travelling between specified points on the pathways in the environments. The normalized path that this process produced is a distilled representation of the position data, which allowed us to measure the overlap values that were not obscured by other effects (such as a participant travelling in a straight line during encoding, versus a sinusoidal path in retrieval). See Figure 2 for a visual representation of the normalized paths used in the calculation of the Route Overlap values. This process resulted in 12 values for percent overlap scores per person, 6 of which corresponded to active trials, and six to passive trials. For each participant, the averages were calculated for the active and passive trials separately, resulting in one measure representing performance following active encoding, and another following passive encoding. See the example in Figure 2.

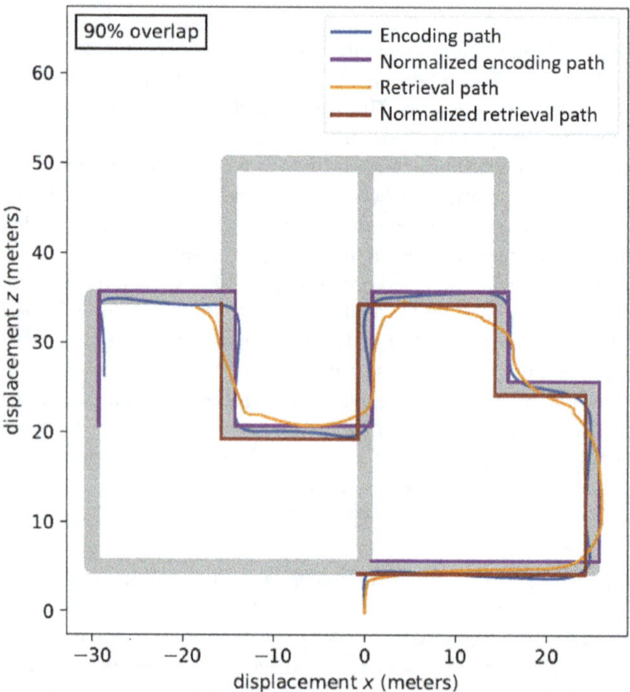

Figure 2. Example of the paths travelled in one of the virtual environments at encoding and retrieval before and after normalization.

Route overlap analyses. A mixed measures ANOVA (Analysis of variance) (2 (Age: younger, older) × 2 (Trial type: active, passive) × 2 (Sex: male, female)) was conducted on the Route Overlap values, with the Trial type as a within-subjects factor, and Age and Sex as between-subjects factors. The analysis revealed a significant main effect of Trial type, $F(1, 40) = 23.07$, $MSE = 0.27$, $p < 0.001$, $\eta^2 = 0.37$, such that the Route Overlap was more accurate for active than passive encoding. A main effect was also found for Age, $F(1, 40) = 83.44$, $MSE = 1.99$, $p < 0.001$, $\eta^2 = 0.68$, such that the Route Overlap was higher for younger than for older adults. No main effect was found for Sex, $F(1, 40) = 0.03$, $MSE = 0.001$, $p = 0.86$, $\eta^2 = 0.06$. There was a significant two-way interaction between Trial type and

Age $F(1, 40) = 5.65$, $MSE = 0.07$, $p = 0.03$, $\eta^2 = 0.12$. The Age $\times$ Sex interactions was non-significant, $F(1, 40) = 2.42$, $MSE = 0.06$, $p = 0.13$, $\eta^2 = 0.06$, as was the Trial type $\times$ Sex, $F(1, 40) = 1.38$, $MSE = 0.02$, $p = 0.25$, $\eta^2 = 0.03$. The 3-way interaction was also non-significant, $F(1, 40) = 0.38$, $MSE = 0.005$, $p = 0.54$, $\eta^2 = 0.01$.

To better understand the interaction between Age and Trial type, two paired-sample *t*-tests were conducted on with Route Overlap accuracy, separately for younger and older adults. For younger adults, there was a marginally significant difference, $t(21) = 1.98$, $p = 0.06$, $d = 0.35$, such that the Route Overlap was higher following active encoding, compared to passive encoding. For older adults, the Route Overlap was significantly higher following active encoding, compared to passive encoding, $t(21) = 4.55$, $p < 0.001$, $d = 0.61$. See Figure 3 for the mean values.

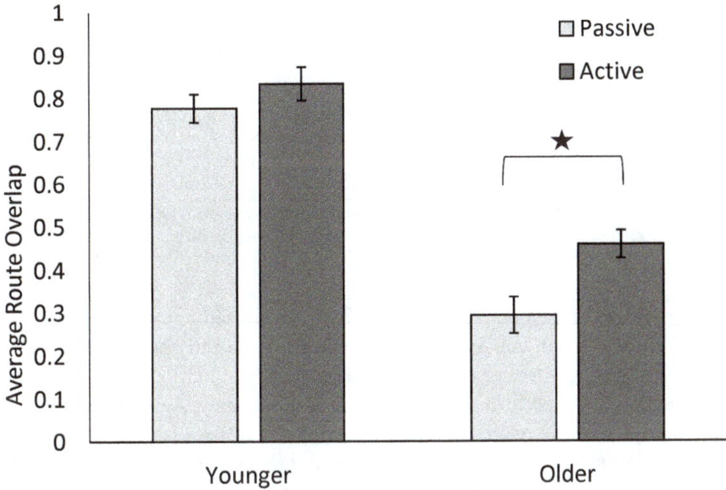

Figure 3. Average percent Route Overlap values for active and passive encoding in younger and older adults. Error bars represent the standard error. The star represents significant differences between active and passive conditions for older adults.

Correlations with neuropsychological assessments. To determine whether spatial memory performance was related to performance in the Hooper Visual Orientation test [45] and the Santa Barbara Sense-of-Direction scale [47], we ran a series of partial correlations separately, for active and passive encoding, while controlling for age. A significant and positive relationship was found between Route Overlap accuracy and the Hooper Visual Orientation test, for both active ($r = 0.51$, $p = 0.001$) and passive ($r = 0.42$, $p = 0.005$) Trial types, suggesting, as expected, that better spatial visualization abilities are related to superior spatial memory performance, regardless of whether one actively or passively encodes a route. Correlations with the Santa Barbara Sense-of-Direction scale were non-significant.

Correlations with cognitive indices. Given our goal of determining the factors underlying the benefits of Active encoding, we conducted exploratory analyses on a number of measures obtained from our virtual reality program. Specifically, a number of partial correlations were analyzed. Table 1 shows the factors considered, along with the correlation and the statistical significance values. Given the large number of correlations, we applied both the Bonferroni [50] and Benjamini–Hochberg corrections [51], the latter of which was included as an alternative to the highly conservative Bonferroni correction. Three correlations remained significant after the Benjamini–Hochberg correction. As the number of stars collected during the active trials increased, so did memory accuracy ($r = 0.39$, $p = 0.009$); this relationship held only for active but not passive encoding trials. During encoding, there were occasions where the participant stopped along the route (during both active and passive trial types); we measured the distance of a stop from the nearest intersection, and found that as this distance got smaller, memory accuracy increased ($r = -0.39$, $p = 0.01$). Similarly, we measured this distance during the retrieval phase (when a participant was re-tracing their route), and found the same relationship ($r = -0.38$, $p = 0.01$). The relationships the distances of the stops from the intersections and Route Overlap held only for the active, and not the passive encoding trials.

Table 1. Correlations between memory accuracy (percent overlap values) following active and passive encoding, and various neuropsychological factors, as well as task performance measures.

	Active	Passive
SBSD	$r = -0.07$, $p = 0.62$	$r = 0.11$, $p = 0.48$
Hooper	$r = 0.51$, $p = 0.001$ [a,b]	$r = 0.42$, $p = 0.005$ [b]
Number of stars passed during encoding	$r = 0.39$, $p = 0.009$ [b]	$r = 0.09$, $p = 0.57$
Number of intersections passed during encoding	$r = -0.006$, $p = 0.97$	$r = 0.10$, $p = 0.52$
Length of the path travelled at encoding	$r = -0.09$, $p = 0.57$	$r = 0.03$, $p = 0.83$
Length of the path travelled at retrieval	$r = -0.004$, $p = 0.98$	$r = -0.13$, $p = 0.41$
Total duration of stopped movement during encoding	$r = 0.06$, $p = 0.72$	$r = 0.03$, $p = 0.86$
Number of stops during encoding	$r = 0.14$, $p = 0.36$	$r = -0.03$, $p = 0.85$
Distance of stops from the nearest intersection during encoding	$r = -0.39$, $p = 0.01$ [b]	$r = -0.01$, $p = 0.96$
Total duration of stopped movement during retrieval	$r = -0.23$, $p = 0.13$	$r = 0.06$, $p = 0.70$
Number of stops during retrieval	$r = 0.04$, $p = 0.82$	$r = 0.01$, $p = 0.95$
Distance of stops from the nearest intersection during retrieval	$r = -0.38$, $p = 0.01$ [b]	$r = -0.25$, $p = 0.11$

Both the Bonferroni and Benjamini–Hochberg corrections for multiple comparisons were conducted; SBSD = Santa Barbara Sense of Direction scale; [a] Correlations were significant following Bonferroni correction; [b] Correlations were significant following Benjamini–Hochberg correction.

Regression analyses. Step-wise regression analyses were used to examine whether any of the factors considered in the correlational analyses (see Table 1) significantly predicted participants' Route Overlap accuracies following active or passive encoding.

In Step 1 of the analysis of performance following active encoding, Age entered into the regression equation, and was significantly related to Route Overlap, $F(1,42) = 75.86$, $p < 0.001$. At Step 2, the Hooper Visual Orientation test performance was entered into the equation, $t = 3.05$, $p = 0.004$, and at Step 3, the distance of participants' stops from the nearest intersection during encoding was entered, $t = -2.18$, $p = 0.035$. At Step 4, the number of stars collected failed to be entered into the equation, $t = 1.94$, $p = 0.06$. The final model significantly predicted Route Overlap, $F(3, 40) = 38.00$, $p < 0.001$, with an R^2 of 0.740, decreasing with Age ($\beta = -0.56$) and distance from the nearest decision point ($\beta = -0.19$), and increasing with the Hooper performance ($\beta = 0.27$).

In Step 1 of the passive encoding analysis, Age entered into the regression equation and was significantly related to Route Overlap, $F(1,42) = 101.82$, $p < 0.001$. At Step 2, the Hooper Visual Orientation test performance was entered into the equation, $t = 2.24$, $p = 0.031$, and at Step 3, the length of the path travelled at retrieval failed to be entered into the equation, $t = -1.60$, $p = 0.12$. The final model significantly predicted Route Overlap, $F(2, 41) = 58.26$, $p < 0.001$, with an R^2 of 0.74, decreasing with Age ($\beta = -0.68$) and increasing with Hooper performance ($\beta = 0.24$).

7. Discussion

In this study we created a series of virtual environments in which younger and older adult participants travelled routes and then later re-traced those routes, allowing us to assess spatial memory performance. Critically, participants either traveled routes during encoding by active exploration through the volitional control of their movements, or by passively viewing a guided tour. As expected, we found that actively navigating through virtual environments resulted in more accurate performances in re-tracing the travelled routes, than passively viewing a recorded navigation during encoding. As expected, those with higher scores on the Hooper Visual Organizational test had more accurate spatial memories, regardless of how the routes were encoded. Importantly, we showed that the memory enhancement from active encoding was larger for older than younger adults, indicating that age-related deficits in spatial memory can be reduced by active encoding. Our analysis also revealed that the cognitive approach of stopping close to an intersection benefitted later spatial memory following active encoding. We suggest that active encoding provides a high degree of environmental support for older adults, by prompting cognitive engagement with the route information, which can serve to improve encoding and subsequent retrieval of that information.

The positive relationship found between memory accuracy and scores on the Hooper Visual Orientation test suggests that individuals who can create better mental visualizations are better at creating a mental image of the travelled routes, which benefits performance when retracing those routes. The Hooper Visual Orientation test measures one's ability to mentally rotate and piece together misarranged visual information to identify the depicted object [45,46]. As such, it may be the case that individuals with higher scores on this test are better at mentally translating and combining the visual components of travelled sections of the route, to create a survey representation of the environment. Indeed, this relationship emerged for both active and passive encoding, which indicates that regardless of how one interacts with an environment, the ability to visualize a mental map supports the memory retrieval of route information.

The finding that older adults' spatial memory performance was enhanced following active relative to passive encoding, is consistent with other findings in the literature [15,25,26]. For example, Sauzéon and colleagues [26] demonstrated that older adults had better memories for objects in a virtual apartment following active exploration rather than a passively guided tour. Additionally, active encoding has been found to benefit older adults' memory for spatial information [15,25] showing that when participants are active 'drivers', their memory is better than when they act as passive 'passengers' in virtual environments. Our findings build on this previous work, by demonstrating that when attempting to retrace a route that is encountered on only one previous occasion, having actively been in control of the initial navigation benefitted memory more than passive guidance. These findings contribute to the literature, suggesting that age-related deficits in spatial memory abilities can be ameliorated through the use of active encoding strategies for navigation.

The finding that memory accuracy was related to the number of stars collected following active, but not passive encoding suggests that during active navigation, attention is indeed oriented towards information that is relevant to the traveled route. Notably, no relationship was found between memory accuracy and the length of the path traveled, meaning that the collection of stars must have provided a unique benefit towards encoding. One possibility is that a memory benefit was obtained through intrinsic motivation or curiosity [35,36] associated with collecting the stars. As described in the introduction, it has been suggested that momentary feelings of curiosity boost memory and executive functions [37], because phasic activation of the noradrenergic and dopaminergic systems modulate hippocampal activity [38]. Regardless of the potential mechanism, it is clear from our results that having a larger number of interesting or notable targets in the environment during encoding enhanced memory recall.

We also found a relationships between mental accuracy and the distance of stops from the nearest intersection during encoding and retrieval, such that stopping near an intersection was related to better performance in retracing the route. Furthermore, in the regression analysis, stopping near an intersection during active encoding was found to be predictive of the accuracy in route overlap. When participants stopped moving near an intersection within the environment during encoding, it facilitated their later memory in recalling the spatial route travelled. It may also be the case that the intersections, or decision points, are critical components of the overall spatial representation of the route. Enhanced encoding of these points likely led to a better spatial representation of the details of the route. When the participants stopped moving in the environment during retrieval, they were possibly attempting to remember the decisions that were previously made at the intersection during encoding, and correcting their planned route if necessary. However, it is important to note that only stopping near an intersection during active encoding, and not retrieval, was a significant predictor of route overlap accuracy. This finding suggests that stopping to focus attention on decision points during exploration in the environment aids spatial memory performance.

In older adults, links have been found between poor spatial representation abilities and declines in both executive functioning [52,53] and episodic memory [27]. The effortfulness hypothesis [54], which states that, because of age-related sensory limitations, there are more cognitive resources that are allocated to the identification of perceptual information, rather than to the encoding of meaningful content, could explain why spatial memory deficits occur in older adults. For example, if limited cognitive resources are directed towards the identification or perceptual processing of a landmark, encoding of the spatial location of the landmark or of the route travelled would be hindered. Indeed, navigational strategies are effortful and cognitively demanding, requiring executive functioning, spatial abilities, and memory [43], which are well-documented as being in decline in older adults. Critically, however, it should be possible to reduce age-related differences in spatial memory performance by implementing encoding strategies that direct cognitive resources specifically to the encoding of spatial information. While our task involved physical components (motor control for locomotion, proprioceptive, and vestibular sensory information) from pressing the movement button and spinning the chair to change directions, it was likely not difficult or demanding enough to create a cognitive burden on the older adults. It would be interesting for future work, to include a divided attention condition, in which participants need to respond to a secondary task while simultaneously navigating during encoding, to determine if heavily taxing their available cognitive resources results in a diminished effect of active encoding.

Notably, we found that older adults had significantly poorer spatial visualization abilities than younger adults, which may be linked to a common deficit that is associated with the poorer spatial memory performance observed in this age group. Older adults have been found to rely more on egocentric rather than allocentric representations during navigation [55,56], the latter of which involves a more highly developed survey representation of space [57], and this is possibly supported by processes that are similar to those supporting general spatial visualization functions. Evidence suggests that degradation in the hippocampal region is related to declines in survey abilities in older adults [58,59] and also to visual perceptual processing [60], which may be involved in spatial scene construction abilities that are supported by the hippocampus [61,62]. Indeed, older adults have been shown to have worse performances than younger adults in both spatial and scene construction tasks that involve imagining an event [63]. An interesting avenue for future work would be to investigate the links between visual imagery abilities, scene construction, and spatial representations, and how these change in the aging brain.

8. Conclusions

The results from this study demonstrate that actively navigating, rather than passively encoding environments results in a more accurate performance when re-tracing a travelled route. Importantly, the magnitude of the memory enhancement from active relative to passive encoding was larger for older than younger adults, reducing the commonly observed age-related deficit in spatial memory performance. Additionally, our findings demonstrate that higher scores on the Hooper Visual Organizational test, and pausing near intersections during active encoding are related to, and predictive of, route overlap accuracy, illuminating the cognitive components that contribute to the benefits of active navigation memory. Overall, our findings suggest that older adults can improve their spatial memory performance for travelled routes by actively controlling their exploration of new environments.

Author Contributions: Conceptualization, M.E.M., M.A.F., and H.S.; methodology, M.E.M., J.G.M., and M.A.F.; software, J.G.M.; validation, M.E.M., J.G.M., and M.A.F.; formal analysis, M.E.M.; investigation, M.E.M.; resources, M.A.F., and H.S.; data curation, M.E.M., and J.G.M.; writing—original draft preparation, M.E.M.; writing—review and editing, M.E.M., J.G.M., M.A.F., and H.S.; visualization, M.E.M. and J.G.M.; supervision, M.A.F.; project administration, M.E.M. and M.A.F.; funding acquisition, M.A.F. and H.S.

Funding: This research was funded by a Natural Sciences and Engineering Research Council of Canada (NSERC) Discovery Grant 2015–2020 awarded to author M.F., and graduate scholarship from NSERC awarded to author M.E.M., as well as a Waterloo-Bordeaux International Initiative grant awarded to M.A.F. and H.S.

Conflicts of Interest: The authors declare no conflict of interest.

References

1. Kirasic, K.C. Spatial cognition and behavior in young and elderly adults: Implications for learning new environments. *Psychol. Aging* **1991**, *6*, 10–18. [CrossRef] [PubMed]
2. Wilkniss, S.M.; Jones, M.G.; Korol, D.L.; Gold, P.E.; Manning, C.A. Age-related differences in an ecologically based study of route learning. *Psychol. Aging* **1997**, *12*, 372–375. [CrossRef] [PubMed]
3. De Beni, R.; Pazzaglia, F.; Gardini, S. The role of mental rotation and age in spatial perspective-taking tasks: When age does not impair perspective-taking performance. *Appl. Cognit. Psychol.* **2006**, *20*, 807–821. [CrossRef]
4. Jonker, C.; Launer, L.J.; Hooijer, C.; Lindeboom, J. Memory Complaints and Memory Impairment in Older Individuals. *J. Am. Geriatr. Soc.* **1996**, *44*, 44–49. [CrossRef] [PubMed]
5. Leitsch, S.; Algase, D.L.; Hong, G.-R.; Beattie, E.; Song, J.-A.; Yao, L. The Interrelatedness of Wandering and Wayfinding in a Community Sample of Persons with Dementia. *Dement. Geriatr. Cogn. Disord.* **2004**, *17*, 231–239.
6. Henderson, V.W.; Mack, W.; Williams, B.W. Spatial Disorientation in Alzheimer's Disease. *Arch. Neurol.* **1989**, *46*, 391–394. [CrossRef] [PubMed]
7. Kavcic, V.; Duffy, C.J. Attentional dynamics and visual perception: mechanisms of spatial disorientation in Alzheimer's disease. *Brain* **2003**, *126*, 1173–1181. [CrossRef] [PubMed]
8. Monacelli, A.M.; Cushman, L.A.; Kavcic, V.; Duffy, C.J. Spatial disorientation in Alzheimer's disease: The remembrance of things passed. *Neurology* **2003**, *61*, 1491–1497. [CrossRef] [PubMed]
9. Pai, M.-C.; Pai, M.; Jacobs, W.J. Topographical disorientation in community-residing patients with Alzheimer's disease. *Int. J. Geriatr. Psychiatry* **2004**, *19*, 250–255. [CrossRef] [PubMed]
10. Cogné, M.; Taillade, M.; N'Kaoua, B.; Tarruella, A.; Klinger, E.; Larrue, F.; Sauzéon, H.; Joseph, P.-A.; Sorita, E. The contribution of virtual reality to the diagnosis of spatial navigation disorders and to the study of the role of navigational aids: A systematic literature review. *Ann. Phys. Rehabil. Med.* **2017**, *60*, 164–176. [CrossRef] [PubMed]
11. Chrastil, E.R.; Warren, W.H. Active and passive contributions to spatial learning. *Psychon. Bull. Rev.* **2011**, *19*, 1–23. [CrossRef] [PubMed]
12. Markant, D.B.; Ruggeri, A.; Gureckis, T.M.; Xu, F. Enhanced Memory as a Common Effect of Active Learning. *Mind Brain Educ.* **2016**, *10*, 142–152. [CrossRef]
13. Chrastil, E.R.; Warren, W.H. Active and passive spatial learning in human navigation: Acquisition of survey knowledge. *J. Exp. Psychol. Learn. Mem. Cognit.* **2013**, *39*, 1520–1537. [CrossRef] [PubMed]
14. Vidal, M.; Kemeny, A.; Gaunet, F.; Berthoz, A. Active, passive and snapshot exploration in a virtual environment: influence on scene memory, reorientation and path memory. *Cogn. Brain Res.* **2001**, *11*, 409–420.
15. Jebara, N.; Orriols, E.; Zaoui, M.; Berthoz, A.; Piolino, P. Effects of Enactment in Episodic Memory: A Pilot Virtual Reality Study with Young and Elderly Adults. *Front. Neuroinform.* **2014**, *6*, 338. [CrossRef] [PubMed]
16. Sandamas, G.; Foreman, N. Active Versus Passive Acquisition of Spatial Knowledge While Controlling a Vehicle in a Virtual Urban Space in Drivers and Non-Drivers. *SAGE Open* **2015**, *5*, 215824401559544. [CrossRef]
17. Brooks, B.M. The Specificity of Memory Enhancement During Interaction with a Virtual Environment. *Memory* **1999**, *7*, 65–78. [CrossRef] [PubMed]
18. Sauzéon, H.; Pala, P.A.; Larrue, F.; Wallet, G.; Déjos, M.; Zheng, X.; Guitton, P.; N'Kaoua, B. The Use of Virtual Reality for Episodic Memory Assessment. *Exp. Psychol.* **2012**, *59*, 99–108. [CrossRef] [PubMed]
19. Zimmer, H.D.; Engelkamp, J. Levels-of-processing effects in subject-performed tasks. *Mem. Cogn.* **1999**, *27*, 907–914. [CrossRef]
20. Karlsson, T.; Backman, L.; Herlitz, A.; Nilsson, L.G.; Winblad, B.; Osterlind, P.O. Memory improvement at different stages of Alzheimer's disease. *Neuropsychologia* **1989**, *27*, 737–742. [CrossRef]
21. Silva, A.R.; Pinho, M.S.; Souchay, C.; Moulin, C.J.A. Evaluating the subject-performed task effect in healthy older adults: Relationship with neuropsychological tests. *Socioaffect. Neurosci. Psychol.* **2015**, *5*, 53. [CrossRef] [PubMed]

22. Essen, J.D.; Von Essen, J.D.; Nilsson, L.-G. Memory effects of motor activation in subject-performed tasks and sign language. *Psychon. Bull. Rev.* **2003**, *10*, 445–449. [CrossRef]
23. Craik, F.I.M.; Routh, D.A.; Broadbent, D.E. On the Transfer of Information from Temporary to Permanent Memory [and Discussion]. *Philos. Trans. R. Soc. B* **1983**, *302*, 341–359. [CrossRef]
24. Craik, F.I.M.; Jennings, J.M. Human memory. In *The Handbook of Aging and Cognition*; Craik, F.I.M., Salthouse, T.A., Eds.; Lawrence Erlbaum Associates: Hillsdale, NJ, USA, 1992; pp. 51–110.
25. Plancher, G.; Tirard, A.; Gyselinck, V.; Nicolas, S.; Piolino, P. Using virtual reality to characterize episodic memory profiles in amnestic mild cognitive impairment and Alzheimer's disease: Influence of active and passive encoding. *Neuropsychologia* **2012**, *50*, 592–602. [CrossRef] [PubMed]
26. Sauzéon, H.; N'Kaoua, B.; Pala, P.A.; Taillade, M.; Guitton, P. Age and active navigation effects on episodic memory: A virtual reality study. *Br. J. Psychol.* **2015**, *107*, 72–94. [CrossRef] [PubMed]
27. Taillade, M.; Sauzéon, H.; Dejos, M.; Pala, P.A.; Larrue, F.; Wallet, G.; Gross, C.; N'Kaoua, B. Executive and memory correlates of age-related differences in wayfinding performances using a virtual reality application. *Aging Neuropsychol. Cognit.* **2013**, *20*, 298–319. [CrossRef] [PubMed]
28. Li, K.Z.; Lindenberger, U.; Freund, A.M.; Baltes, P.B. Walking While Memorizing: Age-Related Differences in Compensatory Behavior. *Psychol. Sci.* **2001**, *12*, 230–237. [CrossRef] [PubMed]
29. Hausdorff, J.M.; Yogev, G.; Springer, S.; Simon, E.S.; Giladi, N. Walking is more like catching than tapping: Gait in the elderly as a complex cognitive task. *Exp. Brain Res.* **2005**, *164*, 541–548. [CrossRef] [PubMed]
30. Bakdash, J.Z.; Linkenauger, S.A.; Proffitt, D. Comparing Decision-Making and Control for Learning a Virtual Environment: Backseat Drivers Learn Where They are Going. *Proc. Hum. Factors Ergon. Soc. Ann. Meet.* **2008**, *52*, 2117–2121. [CrossRef]
31. Bakdash, J.; Linkenauger, S.; Stefanucci, J.; Witt, J.; Banton, T.; Proffitt, D. Perceived distance influences simulated walking time. *J. Vis.* **2010**, *8*, 1151. [CrossRef]
32. Feyereisen, P. Enactment effects and integration processes in younger and older adults' memory for actions. *Memory* **2009**, *17*, 374–385. [CrossRef] [PubMed]
33. Gutchess, A.H.; Kensinger, E.A.; Yoon, C.; Schacter, D.L. Ageing and the self-reference effect in memory. *Memory* **2007**, *15*, 822–837. [CrossRef] [PubMed]
34. Lalanne, J.; Rozenberg, J.; Grolleau, P.; Piolino, P. The Self-Reference Effect on Episodic Memory Recollection in Young and Older Adults and Alzheimer's Disease. *Curr. Alzheimer Res.* **2013**, *10*, 1107–1117. [CrossRef] [PubMed]
35. Ryan, R.M.; Deci, E.L. Intrinsic and Extrinsic Motivations: Classic Definitions and New Directions. *Contemp. Educ. Psychol.* **2000**, *25*, 54–67. [CrossRef] [PubMed]
36. Kashdan, T.B.; Silvia, P.J. Curiosity and Interest: The Benefits of Thriving on Novelty and Challenge. *Curiosit. Interest Benefits Thriving Nov. Chall.* **2012**, *2*, 367–374.
37. Sakaki, M.; Yagi, A.; Murayama, K. Curiosity in old age: A possible key to achieving adaptive aging. *Neurosci. Biobehav. Rev.* **2018**, *88*, 106–116. [CrossRef] [PubMed]
38. Oudeyer, P.Y.; Gottlieb, J.; Lopes, M. Intrinsic motivation, curiosity, and learning: Theory and applications in educational technologies. In *Progress in Brain Research*; Elsevier: Amsterdam, The Netherlands, 2016; Volume 229, pp. 257–284.
39. Hegarty, M.; Montello, D.R.; Richardson, A.E.; Ishikawa, T.; Lovelace, K. Spatial abilities at different scales: Individual differences in aptitude-test performance and spatial-layout learning. *Intelligence* **2006**, *34*, 151–176. [CrossRef]
40. Moffat, S. Age differences in spatial memory in a virtual environment navigation task. *Neurobiol. Aging* **2001**, *22*, 787–796. [CrossRef]
41. Moffat, S.D.; Resnick, S.M. Effects of age on virtual environment place navigation and allocentric cognitive mapping. *Behav. Neurosci.* **2002**, *116*, 851–859. [CrossRef] [PubMed]
42. Cushman, L.A.; Stein, K.; Duffy, C.J. Detecting navigational deficits in cognitive aging and Alzheimer disease using virtual reality. *Neurology* **2008**, *71*, 888–895. [CrossRef] [PubMed]
43. Taillade, M.; N'Kaoua, B.; Sauzéon, H. Age-Related Differences and Cognitive Correlates of Self-Reported and Direct Navigation Performance: The Effect of Real and Virtual Test Conditions Manipulation. *Front. Psychiatry* **2016**, *6*, 125. [CrossRef] [PubMed]

44. Nasreddine, Z.S.; Phillips, N.A.; Bedirian, V.; Charbonneau, S.; Whitehead, V.; Collin, I.; Cummings, J.L.; Chertkow, H.; Bédirian, V. The Montreal Cognitive Assessment, MoCA: A Brief Screening Tool for Mild Cognitive Impairment. *J. Am. Geriatr. Soc.* **2005**, *53*, 695–699. [CrossRef] [PubMed]

45. Buttaro, M. *Hooper Visual Organization Test*; Springer Nature: Basingstoke, UK, 2017; Volume 28, pp. 1–3.

46. Lopez, M.N.; Lazar, M.D.; Oh, S. Psychometric Properties of the Hooper Visual Organization Test. *Assessment* **2003**, *10*, 66–70. [CrossRef] [PubMed]

47. Hegarty, M. Development of a self-report measure of environmental spatial ability. *Intelligence* **2002**, *30*, 425–447. [CrossRef]

48. Brown, J. Some tests of the decay theory of immediate memory. *Q. J. Exp. Psychol.* **1958**, *10*, 12–21. [CrossRef]

49. Moffat, S.D.; Hampson, E.; Hatzipantelis, M. Navigation in a "Virtual" Maze: Sex Differences and Correlation with Psychometric Measures of Spatial Ability in Humans. *Evol. Hum. Behav.* **1998**, *19*, 73–87. [CrossRef]

50. Miceli, A.; Salkind, N. Teoria Statistica Delle Classi e Calcolo Delle Probabilità. In *Encyclopedia of Research Design*; SAGE Publications: Thousand Oaks, CA, USA, 2012.

51. Benjamini, Y.; Hochberg, Y. Controlling the False Discovery Rate: A Practical and Powerful Approach to Multiple Testing. *J. R. Stat. Soc. Ser. B* **1995**, *57*, 289–300. [CrossRef]

52. Ciaramelli, E. The role of ventromedial prefrontal cortex in navigation: A case of impaired wayfinding and rehabilitation. *Neuropsychologia* **2008**, *46*, 2099–2105. [CrossRef] [PubMed]

53. Moffat, S.D.; Kennedy, K.M.; Rodrigue, K.M.; Raz, N. Extrahippocampal Contributions to Age Differences in Human Spatial Navigation. *Cereb. Cortex* **2006**, *17*, 1274–1282. [CrossRef] [PubMed]

54. McCoy, S.L.; Tun, P.A.; Cox, L.C.; Colangelo, M.; Stewart, R.A.; Wingfield, A. Hearing loss and perceptual effort: downstream effects on older adults' memory for speech. *Q. J. Exp. Psychol. Sec. A* **2005**, *58*, 22–33. [CrossRef] [PubMed]

55. Gazova, I.; Laczo, J.; Rubinova, E.; Mokrisova, I.; Hyncicova, E.; Andel, R.; Vyhnalek, M.; Sheardova, K.; Coulson, E.J.; Hort, J. Spatial navigation in young versus older adults. *Front. Neuroinform.* **2013**, *5*, 94. [CrossRef] [PubMed]

56. Serino, S.; Cipresso, P.; Morganti, F.; Riva, G. The role of egocentric and allocentric abilities in Alzheimer's disease: A systematic review. *Ageing Res. Rev.* **2014**, *16*, 32–44. [CrossRef] [PubMed]

57. Siegel, A.W.; White, S.H. The Development of Spatial Representations of Large-Scale Environments. *Adv. Child Dev. Behav.* **1975**, *10*, 9–55. [PubMed]

58. Guderian, S.; Dzieciol, A.M.; Gadian, D.G.; Jentschke, S.; Doeller, C.F.; Burgess, N.; Mishkin, M.; Vargha-Khadem, F. Hippocampal Volume Reduction in Humans Predicts Impaired Allocentric Spatial Memory in Virtual-Reality Navigation. *J. Neurosci.* **2015**, *35*, 14123–14131. [CrossRef] [PubMed]

59. Fidalgo, C.; Martin, C.B. The Hippocampus Contributes to Allocentric Spatial Memory through Coherent Scene Representations. *J. Neurosci.* **2016**, *36*, 2555–2557. [CrossRef] [PubMed]

60. Graham, K.S.; Barense, M.D.; Lee, A.C. Going beyond LTM in the MTL: A synthesis of neuropsychological and neuroimaging findings on the role of the medial temporal lobe in memory and perception. *Neuropsychologia* **2010**, *48*, 831–853. [CrossRef] [PubMed]

61. Schacter, D.L.; Addis, D.R.; Hassabis, D.; Martin, V.C.; Spreng, R.N.; Szpunar, K.K. The Future of Memory: Remembering, Imagining, and the Brain. *Neuron* **2012**, *76*, 677–694. [CrossRef] [PubMed]

62. Maguire, E.A.; Mullally, S.L. The hippocampus: A manifesto for change. *J. Exp. Psychol. Gen.* **2013**, *142*, 1180–1189. [CrossRef] [PubMed]

63. Rendell, P.G.; Bailey, P.E.; Henry, J.D.; Phillips, L.H.; Gaskin, S.; Kliegel, M. Older adults have greater difficulty imagining future rather than atemporal experiences. *Psychol. Aging* **2012**, *27*, 1089–1098. [CrossRef] [PubMed]

Article

Investigating Age-Related Neural Compensation During Emotion Perception Using Electroencephalography

Tao Yang [1,2,*], **Caroline Di Bernardi Luft** [3], **Pei Sun** [1], **Joydeep Bhattacharya** [2] and **Michael J. Banissy** [2]

1. Department of Psychology, Tsinghua University, Beijing 100084, China; peisun@tsinghua.edu.cn
2. Department of Psychology, Goldsmiths, University of London, London SE14 6NW, UK; j.bhattacharya@gold.ac.uk (J.B.); m.banissy@gold.ac.uk (M.J.B.)
3. School of Biological and Chemical Sciences, Queen Mary University of London, London E1 4NS, UK; luft@qmul.ac.uk
* Correspondence: t_yang@tsinghua.edu.cn

Received: 26 December 2019; Accepted: 21 January 2020; Published: 23 January 2020

Abstract: Previous research suggests declines in emotion perception in older as compared to younger adults, but the underlying neural mechanisms remain unclear. Here, we address this by investigating how "face-age" and "face emotion intensity" affect both younger and older participants' behavioural and neural responses using event-related potentials (ERPs). Sixteen young and fifteen older adults viewed and judged the emotion type of facial images with old or young face-age and with high- or low- emotion intensities while EEG was recorded. The ERP results revealed that young and older participants exhibited significant ERP differences in two neural clusters: the left frontal and centromedial regions (100–200 ms stimulus onset) and frontal region (250–900 ms) when perceiving neutral faces. Older participants also exhibited significantly higher ERPs within these two neural clusters during anger and happiness emotion perceptual tasks. However, while this pattern of activity supported neutral emotion processing, it was not sufficient to support the effective processing of facial expressions of anger and happiness as older adults showed reductions in performance when perceiving these emotions. These age-related changes are consistent with theoretical models of age-related changes in neurocognitive abilities and may reflect a general age-related cognitive neural compensation in older adults, rather than a specific emotion-processing neural compensation.

Keywords: emotion perception; ageing; neural compensation; electroencephalogram

1. Introduction

The perception of emotional facial expression plays an important role in interpersonal communication [1,2]. Emotional expression can alter the meaning of speech and the ability to accurately identify emotional content is particularly important in social interactions [2]. Difficulties with emotion perception are associated with specific types of social impairment, including poor interpersonal interaction, reduced social competence and loneliness (e.g., [3,4]). These outcomes can have negative impacts on health and wellbeing (e.g., [5]), thus establishing how the capacity for emotion perception is affected as a function of normal adult aging is an important challenge.

Most studies have consistently demonstrated that older adults appear to have declined perception of some negative facial expressions of emotions, such as anger, sadness, fear and surprise (e.g., [6–10]). In contrast, age-related decline in the perception of happiness is less consistent: early studies suggested similar performance levels for young and older adults when using prototypical emotions [8,11,12] (see [1] for a review), but recent work suggests that there may be differences when more subtle emotional stimuli are used [13].

In addition, one issue involved in previous studies comparing both young and older people's face perceptual abilities is the other-age effect—participants tend to show superior performance in the perception of own versus other age faces [14–17]. Therefore, the decline in the performance of older adults might be due to the use of young adult actors in the task, which favours young adult participants. It is also known that irrespective of the age of the perceiver, processing different age faces expressing the emotion can also influence emotion perception more broadly. For instance, in some studies, emotion perceptual accuracy of older faces was reported to be lower than young faces, as the wrinkles and folds of older faces can reduce the signal clarity of facial expressions [18–20]. Further, some researchers proposed that smile/happiness was easier to be perceived in young faces than older faces (e.g., [21]), whereas decoding facial expression of anger was higher in older faces than younger faces [22].

Age-Related Neural Compensation in Emotion Perception

Functional neuroimaging studies have found that older participants exhibit a different neural activation pattern from younger participants during emotion perception [23–26]. Generally, older people tend to recruit more frontal cortical regions during emotion perception. For example, Gunning-Dixon, Gur, and Perkins [23] investigated age-related neural activations in cortical and limbic regions using functional magnetic resonance imaging (fMRI) by presenting facial stimuli of a mixture of negative emotions to both young and older participants. They observed neural activation in the amygdala and surrounding temporolimbic regions in younger participants, whereas older participants showed neural activations in left inferior and middle frontal regions. Similar results were found by Tessitore and colleagues [24] who compared older and younger participants' neural processing of fearful and threatening stimuli using fMRI and found older participants were associated with increased prefrontal cortical neural responses and lower neural responses in the amygdala and posterior fusiform gyri relative to younger adults.

These studies have uncovered important age-related neural correlates of emotion processing; however, there are limitations. Firstly, most neuroimaging studies have typically investigated facial displays of only negative emotions [23–26]; therefore, it is not entirely clear whether the age-related compensation patterns reflect older people's general emotion processing or processing of negative emotions only. Second, perceptual biases may have influenced previous neuroimaging results, since most have only used younger faces, which may lead to the other-age effect influencing performance [14,17]. Another set of issues that have largely been overlooked in most previous studies is the variation of emotional intensity or task difficulty. Previous studies have found that older adults have more difficulty than younger adults at labelling facial emotion with less intensity [27]. Further, recent work has suggested that age-related differences in emotion matching might be limited to low-intensity expressions [28]. However, high-intensity facial emotions displays that are typically linked with high performance have been used in most neuroimaging studies. This may potentially mask group differences in performance (e.g., see [13]), which leads to two issues: (1) it is unclear how older people's perception of subtle/low-intensity facial expressions are linked with brain activity; and (2) it is unclear how task difficulty/face emotion intensity interacts with age-related neural activation changes.

In addition, much prior work using neuroimaging to investigate emotion processing in ageing has used fMRI [23–36]. Electroencephalography (EEG) has better temporal resolution than fMRI and enables inference about the neural time course of emotional facial expression processing. However, most previous ERP studies focused on studying young people's ERPs during perceiving emotions (e.g., [20,29]). Those ERP studies that have compared young versus older neural activation differences mostly investigated neural activations within specific emotion-related ERP components that were previously defined by studying younger people (e.g., early posterior negativity (EPN)). For example, in one study, the late positive potentials (LPP) amplitudes were investigated as a function of age (between 18–81 years) and revealed that the LPP for negative images decreases linearly with age, but LPP for positive images is age-invariant across most of the adult life span [31]. Wieser et al. [32]

compared young and older people's early posterior negativity (EPN, an index of early emotion discrimination) during emotion (neutral, positive and negative) discrimination tasks. They found that EPN (168–232 ms) was delayed in older adults compared to younger adults, but a later component (232–296 ms) was unaffected. This study suggested an age-related delay of early visual emotion discrimination, but the delay does not seem to influence the discrimination of emotional stimuli. Tsolaki et al. [33] compared older and younger adults' neural responses within face-selective ERP component N170 during viewings of facial images of anger and fear. They found that the N170 early component was modulated by the type of emotions and tends to have larger amplitudes in the elderly than the young during the viewing of negative stimuli (such as "anger" and "fear"). Although these studies have revealed important age-related neural activation differences during emotion processing, it is still not clear: (1) whether older people also recruit additional frontal neural regions during the perception of neutral emotion, and (2) what the effects of face emotion intensity and face-age are in modulating the neural compensation patterns.

With these in mind, the aim of this study is to compare older and young adults' neural activation patterns and behavioral responses during the perception of both low- and high- emotional intensities of anger and happiness facial emotions. In particular, the present study seeks to compare older and young people's compensation patterns at neutral (baseline), easy (high emotional intensities) and hard (low emotional intensities) conditions of anger and happiness emotion perception. By doing so, we seek to learn more about older adults' neural compensation patterns in processing positive (happiness) and negative (anger) facial emotions with different task difficulties, which could speak to contemporary theoretical models. One model of interest is the compensation-related utilization of the neural circuits hypothesis (CRUNCH) model [37]. This model suggests that older people's processing inefficiencies lead the ageing brain to recruit additional neural resources to perform at the level of younger adults: at lower levels of task demand, older adults exhibit a region-specific neural overactivation pattern but they can achieve similar or equivalent behavioural performance as younger adults (successful neural compensation); however, beyond a certain level of task demand, the older adult brain falls short of sufficient neural activation and their behavioural performance declines compared to younger people (neural compensation failure). Based on the predictions of the CRUNCH model, we expect to find that different task demands (face emotion images with low- and high- emotional intensities) might require different levels of neural compensation in older adults.

2. Materials and Methods

2.1. Participants

Thirty-one adult human volunteers participated in our study. Participants were divided into two groups: younger adults ($N = 16$, 12 female, mean ± SD of age: $24 ± 6$ years), and older adults ($N = 15$, 12 female, mean ± SD of age: $69 ± 9$ years). The study was conducted in the U.K., and all participants were native English speakers, with no known history of neurological problems, dyslexia or other language-related problems. They had normal or corrected-to-normal vision (self-reported). They gave written informed consent prior to beginning the experiment. The experimental protocol was approved by the local ethics committee of the Department of Psychology at Goldsmiths.

Level of education, premorbid intelligence (National Adult Reading Test, NART) [34], handedness, screening tests of working memory (digit span) [35], and alexithymia trait [36] were recorded for all participants. Further, we also administered the Mini-Mental State Examination (MMSE) as screening for possible dementia in the older adults group [38]; all participants had scores higher than 24, a usual cut-off limit for dementia [39].

2.2. Stimuli

The stimuli used in this study consisted of computer-generated human faces created by the FaceGen software ((www.facegen.com/products.htm), [40]), which were used in previous studies on emotion perception [18,41,42]. The facial stimuli had no hair or facial hair to avoid gender cues other than facial structure and features. All images were three-dimensional, greyscale, front profile Caucasian faces. Faces of thirty young (18–40 years; 15 female) and thirty old (65 years and above; 15 female) individuals were used. Each facial stimulus displayed anger or happiness in five emotional intensities: 0%, 15%, 30%, 60% and 75%. For both angry and happy faces, we formed two groups: faces with high intensities (60% and 75%), faces with low intensities (15% and 30%), corresponding to easy and difficult trials, respectively (see Figure 1 for examples). Faces with zero intensity correspond to neutral trials. Therefore, we had ten conditions in total (young/old face × angry/happy emotion × low/high intensity, and young/old face × neutral). Each condition comprised 60 experimental trials.

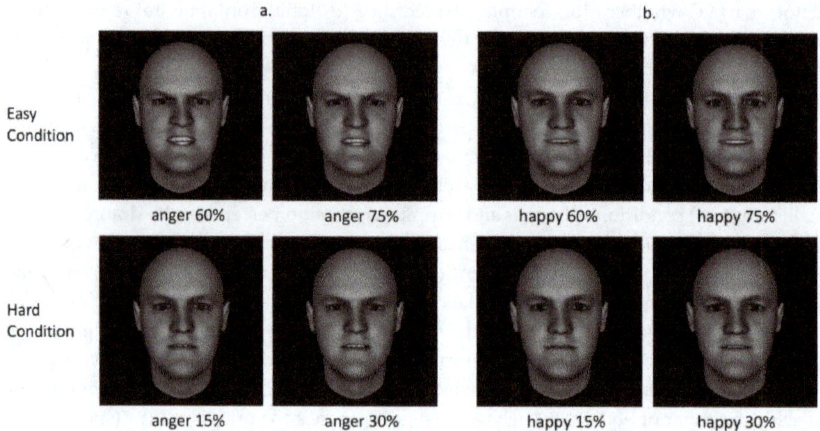

Figure 1. Example of emotional facial stimuli of anger (**a**) and happiness (**b**) used in the experiment. Hard task conditions contain facial stimuli with lower emotional intensities (15% and 30%), and easy task conditions contain facial stimuli with higher emotional intensities (60% and 75%).

2.3. Experimental Procedure

Participants were seated approximately 50 cm in front of a computer screen in a dimly lit, sound-attenuated room, and completed an emotion perception task. In each trial (Figure 2), a face was briefly presented for 500 ms, followed by a fixation cross for 500 ms, and then participants reported emotional category of the earlier face by pressing a key out of three options (anger, happiness, and neutral). The next trial started 1500 ms after the response (see Figure 2). At the beginning of the experiment, participants were informed that some facial expressions were subtle. There was a practice block (30 trials) before the experiment.

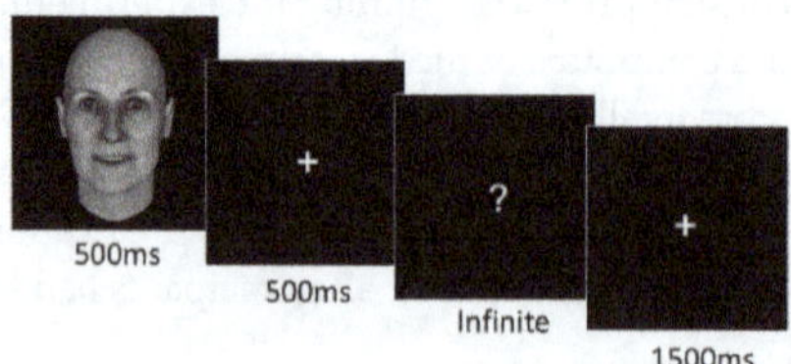

Figure 2. An outline of a trial. In each trial, a face was presented for 500 ms, followed by a blank screen with a fixation cross for 500 ms, and then followed by a prompt, asking participants to provide a response on the emotion category, happy or angry or neutral. The inter-trial interval was 1500 ms.

2.4. EEG Recording and Analysis

EEG signals were recorded with sixty-four Ag-AgCl electrodes placed according to the extended 10–20 electrode placement system, and amplified by a BioSemi ActiveTwo amplifier (www.biosemi.com). Four additional electrodes were placed above and below each eye, and at the outer canthus of each eye, to record vertical and horizontal eye movements, respectively. The sixty-four EEG signals were recorded with a sampling frequency of 512 Hz, bandpass filtered between 0.16–100 Hz. For the sixty-four EEG signals, MATLAB toolbox EEGLAB [43] was used for data preprocessing, and FieldTrip [44] for data analysis and statistical comparisons.

2.5. Preprocessing

The EEG data were algebraically re-referenced to a common average reference for ERP analyses because it is a less biased reference for comparing across scalp topographies [45,46]. The continuous data were high-pass filtered at 0.5 Hz and subsequently epoched from −1500 to 1500 ms time-locked to the onset of a face stimulus. The artifacts were treated in a semi-automated fashion. Visual inspection was initially made to remove any large artifacts, followed by an independent component analysis for correcting the eye blink-related artifacts. Subsequently, epochs containing amplitudes exceeding ±75 µV were discarded from future analysis. Each condition comprised of 60 trials; after artifact rejection, each condition had on average 42 trials remaining. The signals were further low-pass filtered at 35 Hz and epoched from −200 to 1000 ms following the stimulus presentation. The EEG signal of the epochs were averaged to obtain the ERP signals per condition. The ERPs were subsequently baseline corrected (−200 to 0 ms).

2.6. ERP Analysis

To identify the regions of interest (ROIs) on the Electrode x Time Space in the ERP differences between groups (old vs. young) while perceiving happy or angry emotion, we used a data-driven exploratory approach using a neutral emotion condition as a baseline. The ERPs of older and younger adults while perceiving neutral faces were compared by a non-parametric cluster-based permutation test [47], which is hypothesis-free, that controls the Type I error rate, and is widely used to analyse multidimensional EEG/MEG data [48–50]. The method consists of two steps as follows. First, the multidimensional (time, amplitude and electrode) clusters were detected by grouping neighboring data points that showed a significant effect ($p < 0.05$) of group (young vs. old) by independent t-tests, and a cluster-level statistic was subsequently calculated by summing the t-values in each cluster. Here, we considered electrodes with a distance of less than 5 cm as neighbors, yielding on average 4.2 neighbors per electrode. Second, Monte Carlo randomisation was used to calculate the exact probability that a cluster with the maximum cluster-level statistic was observed under the assumption that the ERP values of the two compared groups or conditions were not significantly different. A histogram of maximum cluster-level statistics was obtained by calculating the cluster-level statistic by repeating the entire analysis a large number of times (=500 in the current study) on random permutation of the pooled data of the two groups or conditions; this histogram was subsequently used to calculate the exact p-value for that cluster, and these procedures were subsequently carried out for the lower ranking cluster-level statistics [48]. The clusters thus obtained were subsequently used as the ROIs to find the main differences between the younger and older adults by standard factorial mixed ANOVA on the mean ERP amplitudes at these ROIs. By choosing a neutral condition as the baseline, this method can show the role of emotion type, task difficulty and face-age in modulating the same spatiotemporal brain profiles. Specifically, it can help revealing the participants' neural changes from baseline to easy and hard emotion conditions, which provides the evidence for validating the age-related neural compensation hypothesis.

3. Results

3.1. Participants' Characteristics

Demographic characteristics of the two groups are listed in Table 1. The two groups did not significantly differ ($p > 0.05$) in level of education, premorbid intelligence (NART), handedness, working memory (digit span), or Toronto Alexithymia Scale (TAS-20).

Table 1. Basic demographic and descriptive characteristics of the two study groups.

	Old (n = 15)	*Young (n = 16)*
Gender (male/female)	3/12	4/12
Age (years)	69.20 (8.809)	24.19 (5.576)
Education (years)	15 (3)	16 (2)
Handedness (right/left)	14/1	15/1
Premorbid IQ (NART)	120.07 (8.430)	115.44 (10.295)
Working memory (digit-span)	105.27 (16.867)	106.63 (12.209)
TAS-20 score	43.00 (7.329)	44.75 (9.692)

3.2. Behavioral Results

Prior to data analysis, accuracy outliers were excluded from further analysis (using a criteria of >3 standard deviations from the mean on any individual variable of interest and significance using Grubb's test). Specifically, two datasets from *older face_neutral* condition, one dataset from *older face_anger_easy* condition, one dataset from *younger face_neutral* condition, one dataset from *younger face_anger_easy* condition, and two datasets from *younger face_happy_easy* condition were identified as outliers and excluded from further analysis. No RT outliers were found. Participants' perceptual performance (accuracy and RTs) on each task were analysed.

3.2.1. Perception of Neutral Facial Expression

Perceptual performance on neutral emotion condition was analysed by a 2 × 2 mixed factorial ANOVA with group (young, old) as a between-participants factor and face-age (young, old) as a within-participants factor. Older adults showed marginally higher accuracy on recognizing neutral emotion than younger adults ($F(1, 26) = 4.195$, $p = 0.051$, $\eta^2 = 0.139$; Figure 3). No other significant main effect or interaction was found (see Supplementary Material).

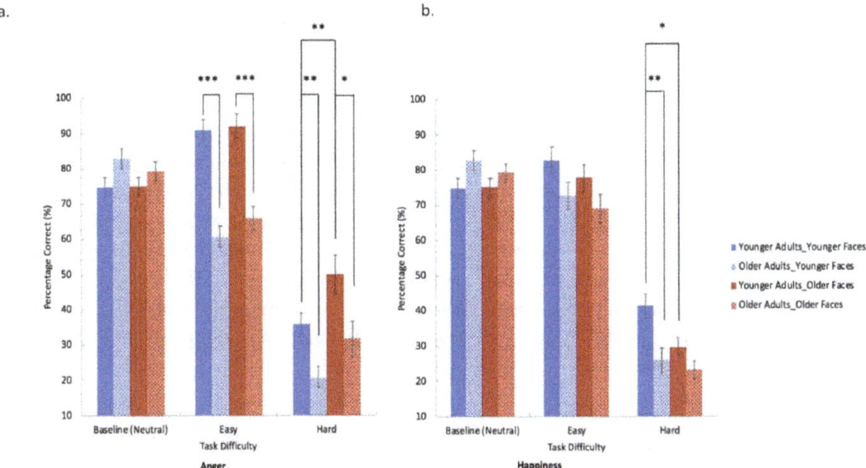

Figure 3. Older (patterned bars) and younger (solid bars) participants' perceptual performance (accuracy) in anger (**a**) and happiness (**b**) experimental trials at different task difficulties (neutral (baseline), easy and hard). Bars in red represent trials with older face stimuli, bars in blue represent trials with younger face stimuli. (1) Anger perception (**a**): older group performed worse than younger group in both easy (old face stimuli condition and young face stimuli condition) and hard (old face stimuli condition and young face stimuli condition) levels. (2) Happiness perception (**b**): older adults performed poorer in hard condition using younger face stimuli, but not in other conditions. (3) In addition, within-group comparison revealed that younger participants' performance on young and old face stimuli were significantly different for only hard level of anger and happiness perception tasks, with superior performance on younger face stimuli in happiness perceptual tasks and superior performance on older face stimuli in anger perceptual tasks. Error bars represents S.E. (*** represents $p < 0.001$. ** represents $p < 0.01$. * represents $p < 0.05$).

3.2.2. Perception of Anger and Happiness Facial Expressions

Perceptual performance on anger and happiness trials were analysed by a 2 × 2 × 2 × 2 mixed ANOVA with group (young, old) as the between-participants factor, and emotion type (anger, happiness), face-age (young, old), and task difficulty (easy, hard) as within-participants factors. Younger adults showed overall better performance than the older adults ($F(1, 26) = 30.357, p < 0.001, \eta^2 = 0.539$; Figure 3). The main effect of task difficulty was also significant ($F(1, 26) = 381.512, p < 0.001, \eta^2 = 0.936$), which was due to the overall accuracy for easy trials being significantly higher than hard trials. The main effects of emotion type and face-age were not significant. The interaction of emotion type × face-age was significant ($F(1, 26) = 24.495, p < 0.001, \eta^2 = 0.485$). Pairwise comparisons (with Bonferroni correction) revealed that this was due to the overall accuracy for older face stimuli being significantly higher than younger face stimuli for anger emotion (anger: $p < 0.001; d = 1.699$). For the happy facial emotion, the overall accuracy for younger face stimuli was significantly higher than for older face stimuli (happiness: $p = 0.008, d = 1.441$, Bonferroni corrected).

The interaction of emotion type × face-age × task difficulty was also significant ($F(1, 26) = 15.085, p = 0.001, \eta^2 = 0.367$). Pairwise comparison (Bonferroni-corrected) revealed that participants' overall accuracy for older face stimuli was significantly higher than younger face stimuli in the hard level of anger perceptual tasks ($p < 0.001, d = 1.744$). In the hard level of happiness perceptual tasks, participants' overall accuracy for younger face stimuli was significantly higher than older face stimuli ($p < 0.001, d = 1.538$).

The interaction of emotion type × face-age × task difficulty × group was also significant ($F(1, 26) = 4.893$, $p = 0.036$, $\eta^2 = 0.158$). Pairwise comparison (Bonferroni-corrected) revealed that the older group performed worse than the younger group at anger perception in both easy (old face stimuli condition: $p < 0.001$, $d = 1.608$; young face stimuli condition: $p < 0.001$, $d = 1.865$) and hard (old face stimuli condition: $p = 0.032$, $d = 1.203$; young face stimuli condition: $p = 0.008$, $d = 1.463$) levels (Figure 3a). Older adults also performed poorer in hard condition of happiness perception using younger face stimuli ($p = 0.008$, $d = 1.443$), but not in other conditions (Figure 3b). In addition, within-group comparison revealed that younger participants' performance on young and old face stimuli were significantly different for only the hard level of anger and happiness perception tasks (anger: $p = 0.008$, $d = 1.438$; happiness: $p < 0.001$, $d = 1.512$), with superior performance on younger face stimuli in happiness perceptual tasks and superior performance on older face stimuli in anger perceptual tasks. In contrast, older participants' performances on younger and older face stimuli were not significantly different in both easy and hard levels of anger and happiness perception tasks.

3.3. ERP Analysis

3.3.1. Selections of ERP Clusters that Showed Difference between Old and Young Group

In order to select the ROIs, we compared young and older participants' ERPs (young vs. old) on neutral emotion condition by using the non-parametric cluster permutation tests (ERPs corresponding to the old and young faces stimuli trials of neural condition were merged). The significant clusters are shown in Figure 4. We identified two significant clusters: the first cluster was a left-frontal and centromedial cluster between 100–200 ms (Figure 4a), and the second cluster was over the frontal region between 250–900 ms (Figure 4b). These two spatiotemporally distinct clusters served as our two ROIs.

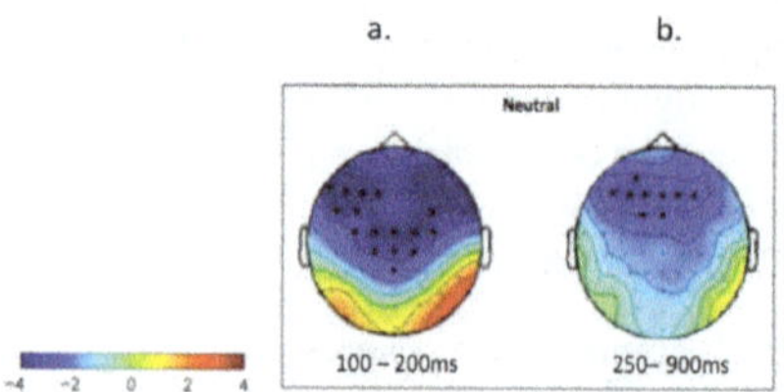

Figure 4. Significant clusters of the nonparametric cluster randomisation test comparing the two groups (young minus old) on neutral condition. First cluster time window: 100–200 ms, left-frontal and centromedial area (F1, F3, F5, F7, Fc5, Fc5, C1, C3, Cp1, Pz, Cpz, Fc4, Cz, C2, C4); second cluster time window: 250–900 ms; frontal area (AF3 F1, F3, F5, Fc1, Fz, F2, F4, Fcz).

3.3.2. The Effect of Emotion, Face-age and Task Difficulty in Modulating ERPs

For each cluster, we computed individual participant mean ERP amplitudes, and the data were subsequently analyzed by a mixed 2 × 2 × 2 × 2 ANOVA with emotions (anger, happiness), face-age (young and old), task difficulty (easy, hard) as within-participants factors, and group (young, old) as a between-participants factor.

Cluster one

The results revealed that older participants exhibited significantly higher overall mean ERP amplitudes than younger group (main effect of group, $F(1, 26) = 20.273$, $p < 0.001$, $\eta^2 = 0.438$; Figure 5). In addition, the easy condition (high-intensity facial expressions) elicited significantly higher overall mean ERP amplitudes in left frontal and centromedial regions (100–200 ms) than the hard condition (low-intensity facial expression) [the main effect of task difficulty, $F(1, 26) = 4.974$, $p = 0.035$, $\eta^2 = 0.161$]. The main effects of emotion type and face-age were not significant.

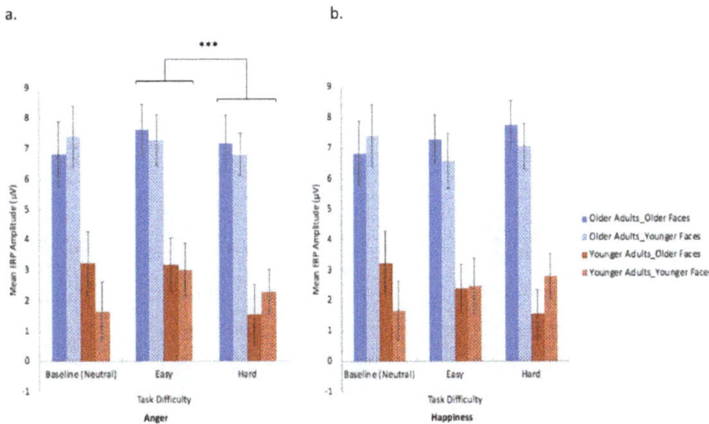

Figure 5. Older (bars in blue) and younger (bars in red) participants' cluster one mean ERP amplitude for anger (**a**) and happiness (**b**) experimental trials at different task difficulties (neutral (baseline), easy and hard). Solid bars represent trials with older face stimuli; patterned bars represent trials with younger face stimuli. The overall left-frontal and centromedial mean ERP amplitude (100–200 ms) of easy condition was significantly higher than the hard condition in anger task, regardless of group. Error bars represents S.E. (*** represents $p < 0.001$).

The interaction of face-age × group was significant [$F(1, 26) = 5.775$, $p = 0.024$, $\eta^2 = 0.182$]. Pairwise comparison (with Bonferroni correction) revealed that older participants exhibited significantly higher mean ERP amplitudes in left frontal and centromedial regions (100–200 ms) than younger participants in both older and younger face stimuli conditions (older face stimuli, $p < 0.001$, $d = 1.764$; younger face stimuli, $p < 0.001$, $d = 1.583$). No significant results was found in within-group comparisons (old group: $p = 0.054$ and young group: $p = 0.172$; with Bonferroni correction; Figure 5). The interaction of emotion type × difficulty was significant ($F(1, 26) = 6.632$, $p = 0.017$, $\eta^2 = 0.201$). Pairwise comparison revealed that the mean left-frontal and centromedial ERP amplitude (100–200 ms) of the easy condition was significantly higher than the hard condition in anger task ($p < 0.001$, $d = 1.537$) (Figure 5).

Cluster two

A 2 (emotion type) × 2 (task difficulty) × 2 (face-age) × 2 (group) mixed ANOVA on the ERP amplitudes of the second cluster (250–900 ms) revealed significant main effect of group ($F(1, 26) = 17.051$, $p < 0.001$, $\eta^2 = 0.396$), which was due to older participants exhibiting significantly higher overall mean ERP amplitudes than younger group (Figure 6). The main effect of task difficulty was significant ($F(1, 26) = 6.813$, $p = 0.015$, $\eta^2 = 0.208$), which was due to the easy condition (high-intensity facial expressions) eliciting significantly higher overall mean ERP amplitudes in frontal regions (250–900 ms) than the hard condition (low-intensity facial expression) (Figure 6). The main effects of emotion type and face-age were not significant.

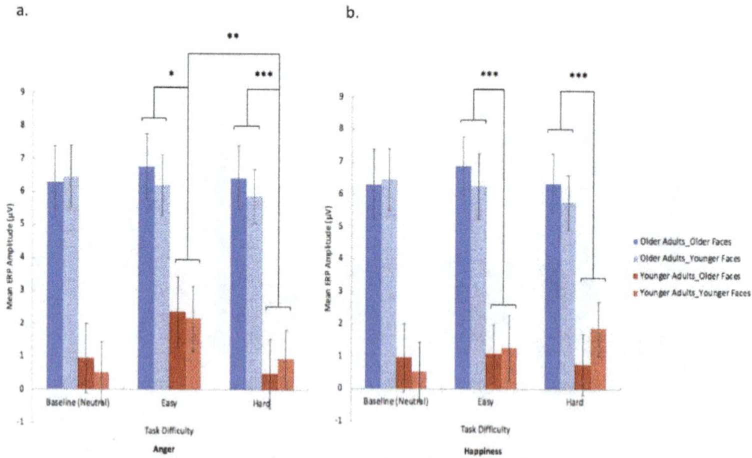

Figure 6. Older (bars in blue) and younger (bars in red) participants' cluster two mean ERP amplitude for anger (**a**) and happiness (**b**) experimental trials at different task difficulties (neutral (baseline), easy and hard). Solid bars represent trials with older face stimuli; patterned bars represent trials with younger face stimuli. (1) Older participants exhibited significantly higher frontal ERP amplitudes (250–900 ms) in both easy and hard levels of anger and happiness perceptual tasks compared to younger participants. (2) Within-group comparison revealed that younger participants' mean frontal ERP amplitudes (250–900 ms) for the easy condition was significantly higher than the hard condition in anger perception, but not significantly different in the happiness perceptual tasks; whereas older participants did not show significant different ERPs across easy and hard task conditions in both the happiness and anger perceptual tasks. Error bars represents S.E. (significances are marked by *** $p < 0.001$, ** $p < 0.01$, and * $p < 0.05$).

The interaction of face-age × group was significant, $F(1, 26) = 5.468$, $p = 0.027$, $\eta^2 = 0.174$, which was due to older adults exhibiting significantly higher ERPs than younger adults in both older ($p < 0.001$, $d = 1.641$) and younger ($p = 0.008$, $d = 1.428$, Bonferroni corrected) face stimuli (Figure 6). In addition, within-group comparison revealed that the older group's mean ERP amplitude for younger faces was significantly lower than for older faces ($p = 0.047$, $d = 0.216$), whereas the younger group's mean ERP for young and old face stimuli did not differ ($p = 0.221$).

The interaction of emotion type × task difficulty × group was significant ($F(1, 26) = 5.781$, $p = 0.024$, $\eta^2 = 0.182$). Pairwise comparison (with Bonferroni correction) revealed that older participants exhibited significantly higher frontal ERP amplitudes (250–900 ms) in both easy and hard levels of anger (easy: $p = 0.048$, $d = 1.367$; hard: $p < 0.001$, $d = 1.766$) and happiness (easy: $p < 0.001$, $d = 1.540$; hard: $p < 0.001$, $d = 1.511$) perceptual tasks. In addition, the within-group comparison revealed that younger participants' mean frontal ERP amplitudes (250–900 ms) for the easy condition was significantly higher than the hard condition in anger perception ($p = 0.008$, $d = 1.487$), but not significantly different in the happiness perceptual tasks, whereas older participants did not show significant different ERPs across easy and hard task conditions in both the happiness and anger perceptual tasks (Figure 6).

4. Discussion

The aim of the present study is to examine age-related changes in emotion perception and to further reveal the underlying neural mechanisms. Our study contributes to the current literature in two main ways. First, most neuroimaging studies have typically investigated people's emotion perception by using facial displays of (1) only negative emotions, (2) only young faces, or (3) only high-emotional intensity. To overcome these limitations, we used an integrative approach by considering how "emotion type", "face-age" and "face emotion intensity" affect younger and older participants' emotion perception, all in the same study. Second, one of the main strengths and novelty of our study is related to our data-driven approach. We used an entirely data-driven approach to explore the age-related neural differences during emotion perception—this is an important contribution as we did not limit ourselves to the known ERP components (e.g., [29,30]). Our study provides additional insights by showing that: (1) older participants exhibited neural differences in the left frontal and centromedial region (100–200 ms stimuli onset) and frontal region (250–900 ms stimuli onset) at neutral condition, which suggests that older people's neural compensation may start at the neutral (baseline) condition; (2) within the same neural clusters, older participants also exhibited similar patterns of neural activity during anger and happiness emotion perceptual tasks; (3) older people's compensation was successful in processing neutral emotion as they exhibited similar behavioural performance as younger participants; however, this was not enough for them to successfully apply the compensation strategy in processing facial expressions of anger and happiness (hard condition). We now discuss these in more detail below.

4.1. Age-Related Behavioural Perceptual Performance Differences

For the perception of anger, older participants' mean accuracy was significantly lower than younger participants in both hard and easy conditions. This finding suggests that older participants show declined abilities in processing both low- and high-intensities of anger facial expressions. This finding adds confirmation to prior work indicating that older participants show declined perception of anger (see [1] for review). For the perception of happiness, older participants showed poorer performance than younger participants in perceiving low-intensity/subtle happiness from younger faces; however, their perception of high-intensity happiness remain intact. This is consistent with previous findings that older participants exhibit significantly poorer performance in perceiving low-intensity happiness from older faces (e.g., [13]), but relatively intact performance on prototypical emotions displayed by younger adults [7–9,51]. Furthermore, it has recently been found that older adults are associated with larger congruency sequence effect [52,53] in the Stroop task compared to younger adult controls [54]. In future studies, it will be interesting to investigate the trial-to-trial carryover effect in emotion perception and to see how typical aging affects this interference effect.

4.2. Effect of Face-Age on Behavioural Perceptual Performance

For anger, younger participants' accuracy for perceiving anger from old face stimuli was significantly higher than from younger faces in the hard condition, which is consistent with Hühnel et al.'s [22] finding that facial expression of anger is easier to detect from older faces than from young faces. This might suggest that some facial features (e.g., winkles around eyes and mouths) of older faces may exaggerate subtle expressions of anger. However, the superior performance in perceiving subtle anger from older faces was not found in older participants, as their performance on younger and older face stimuli was not significantly different in easy or hard levels of anger. It will be important to explore this question further in future work using alternative stimuli to those used in the current design.

For happy emotion, within-group comparison revealed that the younger participants' performance in perceiving low-intensity happiness from younger faces was significantly better than from older faces, which is consistent with Murphy et al.'s finding [21] that smile/happiness is easier to perceive in young than older faces, whereas, older participants' performance in perceiving happiness from younger and older faces did not significantly differ in either low or high intensity. Younger participants' superior performance on younger faces trials (in low emotion intensity condition) might reflect "own-age bias" where people are better at perceiving faces of their own ages [55]. However, this "own-age bias" was not shown in older participants. Younger participants' superior performance in perceiving subtle happiness from younger faces might partially account for the perceptual performance gap between younger and older participants. In other words, the reported poorer performance of older participants might not be entirely due to their perceptual decline, but younger participants' superior perception of their own-age faces.

4.3. *Facial Emotion Perception and Neural Compensation in Older People Revealed by ERPs*

Previous ERP studies have revealed three major facial emotion-related components, which are early frontocentral positivity (around 120 ms stimulus onset), a later broadly distributed sustained positivity beyond 250 ms post-stimulus, and enhanced negativity at lateral posterior sites (EPN) [30,56,57]. The results of the current study revealed that older participants showed significantly higher early frontal and centromedial ERP positivity (100–200 ms) and a later sustained positivity beyond 250 ms post-stimulus across all facial emotions types including neutral emotion.

Based on earlier studies of Eimer et al. [29,30], the early and late frontal and centromedial positivity can be interpreted to reflect a non-automatic and attentive approach of emotion processing, which do not appear to be modulated by emotion type. Older people's significantly higher early (100–200 ms) and late (250–900 ms) frontal and centromedial positivities compared to younger participants during emotion processing may be interpreted as evidence to suggest that older adults require more cortical neural resources than younger people in all types of emotion perception. This additional recruitment of frontal regions during processing facial emotional expression is in parallel with previous fMRI studies, which also found that older people showed significantly higher frontal neural activations compared to younger adults during emotion perception tasks (e.g., [24,25]).

One aim of the present study is to examine the relationship between emotion task difficulty (emotional intensity) and older participants' age-related neural compensation patterns. The ERP results revealed that older participants exhibited significantly higher activations in the left frontal and centromedial regions (100–200 ms stimuli onset) and frontal region (250–900 ms stimuli onset) compared to young participants when perceiving neutral facial expression. This might suggest that neural differences between young and old participants may start at the neutral (baseline) condition in order to compensate for cortical processing inefficiency. In addition, older participants also exhibited similar patterns during perceiving easy- and hard- levels of anger and happiness facial expressions. When combined with the behavioural results, the findings seem to suggest efficient processing of neutral emotion as they performed equally well as younger participants on neutral emotion perceptual trials. A similar level of neural resources was not enough for older adults to successfully process facial expressions of anger (both easy and hard conditions) and happiness (hard condition), where they showed significantly poorer perceptual performance on both happiness (only hard condition) and anger (both easy and hard conditions) perceptual trials. This pattern of results seems in line with the CRUNCH model [37], which proposes that at a lower level of task demand, older participants exhibit a region-specific neural overactivation pattern, but they can achieve similar or equivalent behavioural as younger participants (successful neural compensation). However, beyond a certain level of task demand, there is insufficient neural activation and their behavioural performance declines compared to the young people (neural compensation failure).

It should be noted that the neural overactivation in left frontal and central regions (100–200 ms stimuli onset) and frontal region (250–900 ms stimuli onset) might not reflect emotion-specific neural compensation. Other aging studies have also shown that older people tend to recruit more frontal cortical regions than younger people when performing identical cognitive tasks, especially effortful tasks [58,59]. This has been found in other brain imaging studies, including melodic expectancy processing [60], attention [61–63], working memory [64,65] and executive functioning [66]. Therefore, the neural overactivation in frontal-centro regions might reflect a top-down and controlled neural compensation to suppress or inhibit irrelevant information rather than emotion-specific neural compensation (e.g., [67]). It is also important to note that we only compared the ERP amplitudes of the same clusters (young–old neural activation difference extracted from neutral condition), which might potentially ignore some emotion-specific neural compensation patterns.

4.4. *Effects of "Emotion Type", "Task Difficulty" and "Face-Age" on ERPs*

The results showed that the effect of emotion type (happiness and anger) did not significantly modulate younger and older participants' early (100–200 ms) or late (250–900 ms) positivities during facial emotion perceptual tasks. These findings are in line with previous studies that claimed the early and late positivities are not modulated by emotion type [29,30]. The effect of face-age seems to modulate older people's frontal region ERP between 250–900 ms. Older group's mean ERP amplitude for older faces was significantly higher than for younger faces, which might be due to that the facial expressions portrayed by older faces are harder to be processed compared to those portrayed by younger faces [18–20], whereas younger group mean ERP for young and old face stimuli did not differ, which might suggest that younger people's neural processing for young and older emotional faces is similar and they need to recruit more neural regions to process older emotional faces. This finding is novel and awaits further replication.

In contrast, the effect of task difficulty (or facial emotional intensities) significantly modulated the overall ERP amplitudes of both early (100–200 ms) frontal and centromedial and late (250–900 ms) frontal brain regions in anger tasks, but not in happiness tasks. Specifically, the overall mean left-frontal and centromedial ERP amplitudes (100–200 ms) of easy condition (high emotion intensity) was significantly higher than hard condition (low emotion intensity) in anger task, regardless of group and face-age. In other words, higher intensities of anger elicited significantly higher early (100–200 ms) frontal and centromedial positivities in participants, regardless of group and face-age. Furthermore, younger participants' mean frontal ERP amplitudes (250–900 ms) for the easy condition was significantly higher than the hard condition (low emotion intensity) in anger perception, but this pattern was not shown in the happiness condition. Whereas, older participants did not show significant different ERPs across easy and hard conditions in both happiness and anger perceptual tasks. In other words, higher intensities of anger elicited significantly higher late (250–900 ms) frontal positivities in younger participants only. These findings suggest that higher intensities of anger can trigger higher early (100–200 ms) frontal and centromedial positivities and late (250–900 ms, only in younger participants) frontal positivities. These findings seem consistent with Eimer et al.'s [29,30] interpretation that these two positivity components may contribute to attention-regulated emotional processing, as higher intensities of anger signal potential threat and they are normally associated with a higher level devotion of attention [68]. In addition, a recent study has also shown that anger elicits significantly higher late P3 component (350–450 ms at temporo-occipital and parieto-occipital channel locations) compared to happiness in both younger and older adults [69]. The present finding also partially agrees with a previous proposal suggesting high intensities of emotional images elicited larger late positive potentials than lower intensities of emotional images [70], as it is not the case for happiness perception in the present study.

4.5. Limitations and Future Avenues

Most of the emotion-related ERP components were conceptualized by observing younger adults' ERPs (e.g., [29,30]). Later EEG studies that investigated older people's emotion processing only compared young–old neural differences within these previously defined electrodes and time windows (e.g., [32,33]). However, ageing can lead to older people recruiting additional neural regions during emotion processing (neural compensation), that may extend beyond these already defined EEG neural sites. Recently, some ageing studies started to investigate how ageing affects the previously defined emotion-related components [31–33], but fewer studies have investigated how ageing and other factors (task difficulty, emotion type) affect the ERPs of other neural sites during perceiving facial emotions. In this study, instead of focusing on specific ERP components a priori, we adopted a data-driven approach and selected the ROIs based on the differences between younger and older adults for the neutral condition; therefore, the neutral condition serves as a baseline and helps to reveal the changes in participants' neural responses from baseline to easy and hard emotion conditions. By using the neutral condition, we can examine the effects of emotion type and task difficulty in modulating participants' age-related neural compensation patterns when processing facial expressions, which provides the evidence to demonstrate the age-related neural compensation. There is some evidence that even neutral is processed as an expression [71,72], therefore contrasting the ERP responses between young and older participants on neutral condition gave us the signal differences related to age when trying to recognize emotion in general. In addition, contrasting young and old neural responses to neutral faces enabled us to compare their responses to emotional faces which makes the analyses non-circular to avoid the danger of double-dipping as we tested the differences between the groups on data that we did not use to explore the age-related differences [73]. This is our second reason—as by doing this, we were able to use a data-driven approach to explore the data as a first step (using appropriate methods taking into account the multiple comparison problem) and to test these differences, an unseen dataset when stronger emotions of high and low intensity were being presented.

While useful to address this question, our approach might have ignored the group ERP differences in response to specific emotions, as prior studies have suggested that neutral, anger and happiness facial emotions are processed by different neural regions (e.g., [74,75]). In addition, our selections might have neglected emotion-specific neural compensation in other brain areas, which awaits further neuroimaging studies (e.g., fMRI) to investigate in depth.

Supplementary Materials: The following are available online at http://www.mdpi.com/2076-3425/10/2/61/s1. It include other results and findings that are not included in the main contents [results of RTs, results of accuracy and ERP analysis, results of ERP components P100 and P300].

Author Contributions: T.Y., M.J.B., and J.B. contributed to the study conception and design. T.Y. carried out data collection. Data were analysed and interpreted by C.D.B.L. and T.Y. The drafting of manuscript was written by T.Y. and was critically revised by M.J.B., J.B., C.D.B.L. and P.S. All authors have read and agreed to the published version of the manuscript.

Funding: The project was supported by an ESRC grant awarded to MJB [ES/K00882X/1]. The project was also supported by a grant awarded to PS from the Natural Science Foundation of China [81671065] and a grant awarded to TY from the China Postdoctoral Science Foundation [2018M641311].

Acknowledgments: We would like to thank Amna Ghani, Danying Wang, and Bin Chen for their assistance and support.

Conflicts of Interest: The authors declare that the research was conducted in the absence of any commercial or financial relationships that could be construed as a potential conflict of interest.

References

1. Ruffman, T.; Henry, J.D.; Livingstone, V.; Phillips, L.H. A meta-analytic review of emotion recognition and aging: Implications for neuropsychological models of aging. *Neurosci. Biobehav. Rev.* **2008**, *32*, 863–881. [CrossRef]

2. Ryan, M.; Murray, J.; Ruffman, T. Aging and the perception of emotion: Processing vocal expressions alone and with faces. *Exp. Aging Res.* **2009**, *36*, 1–22. [CrossRef]

3. Spell, L.A.; Frank, E. Recognition of nonverbal communication of affect following traumatic brain injury. *J. Nonverbal Behav.* **2000**, *24*, 285–300. [CrossRef]

4. Kanai, R.; Bahrami, B.; Duchaine, B.; Janik, A.; Banissy, M.J.; Rees, G. Brain structure links loneliness to social perception. *Curr. Biol.* **2012**, *22*, 1975–1979. [CrossRef] [PubMed]

5. Holt-Lunstad, J.; Smith, T.B.; Baker, M.; Harris, T.; Stephenson, D. Loneliness and social isolation as risk factors for mortality: A meta-analytic review. *Perspect. Psychol. Sci.* **2015**, *10*, 227–237. [CrossRef] [PubMed]

6. McDowell, C.L.; Harrison, D.W.; Demaree, H.A. Is right hemisphere decline in the perception of emotion a function of aging? *Int. J. Neurosci.* **1994**, *79*, 1–11. [CrossRef]

7. Phillips, L.H.; MacLean, R.D.; Allen, R. Age and the understanding of emotions neuropsychological and sociocognitive perspectives. *J. Gerontol. Ser. B Psychol. Sci. Soc. Sci.* **2002**, *57*, 526–530. [CrossRef]

8. Calder, A.J.; Keane, J.; Manly, T.; Sprengelmeyer, R.; Scott, S.; Nimmo-Smith, I.; Young, A.W. Facial expression recognition across the adult life span. *Neuropsychologia* **2003**, *41*, 195–202. [CrossRef]

9. Sullivan, S.; Ruffman, T. Emotion recognition deficits in the elderly. *Int. J. Neurosci.* **2004**, *114*, 403–432. [CrossRef]

10. Mill, A.; Allik, J.; Realo, A.; Valk, R. Age-related differences in emotion recognition ability: A cross-sectional study. *Emotion* **2009**, *9*, 619. [CrossRef]

11. Moreno, C.; Borod, J.C.; Welkowitz, J.; Alpert, M. The perception of facial emotion across the adult life span. *Dev. Neuropsychol.* **1993**, *9*, 305–314. [CrossRef]

12. Orgeta, V.; Phillips, L.H. Effects of age and emotional intensity on the recognition of facial emotion. *Exp. Aging Res.* **2007**, *34*, 63–79. [CrossRef] [PubMed]

13. Yang, T.; Penton, T.; Köybaşı, Ş.L.; Banissy, M.J. Social perception and aging: The relationship between aging and the perception of subtle changes in facial happiness and identity. *Acta Psychol.* **2017**, *179*, 23–29. [CrossRef] [PubMed]

14. Anastasi, J.S.; Rhodes, M.G. An own-age bias in face recognition for children and older adults. *Psychon. Bull. Rev.* **2005**, *12*, 1043–1047. [CrossRef] [PubMed]

15. Ebner, N.C.; He, Y.I.; Johnson, M.K. Age and emotion affect how we look at a face: Visual scan patterns differ for own-age versus other-age emotional faces. *Cogn. Emot.* **2011**, *25*, 983–997. [CrossRef] [PubMed]

16. Ebner, N.C.; Johnson, M.K. Young and older emotional faces: Are there age group differences in expression identification and memory? *Emotion* **2009**, *9*, 329. [CrossRef]

17. Wright, D.B.; Stroud, J.N. Age differences in lineup identification accuracy: People are better with their own age. *Law Hum. Behav.* **2002**, *26*, 641. [CrossRef]

18. Hess, U.; Adams, R.B., Jr.; Simard, A.; Stevenson, M.T.; Kleck, R.E. Smiling and sad wrinkles: Age-related changes in the face and the perception of emotions and intentions. *J. Exp. Soc. Psychol.* **2012**, *48*, 1377–1380. [CrossRef]

19. Courgeon, M.; Buisine, S.; Martin, J.C. Impact of expressive wrinkles on perception of a virtual character's facial expressions of emotions. In *International Workshop on Intelligent Virtual Agents*; Springer: Berlin/Heidelberg, Germany, 2009; pp. 201–214.

20. Richter, D.; Dietzel, C.; Kunzmann, U. Age differences in emotion recognition: The task matters. *J. Gerontol. Ser. B Psychol. Sci. Soc. Sci.* **2010**, *66*, 48–55. [CrossRef]

21. Murphy, N.A.; Isaacowitz, D.M. Age effects and gaze patterns in recognising emotional expressions: An in-depth look at gaze measures and covariates. *Cogn. Emot.* **2010**, *24*, 436–452. [CrossRef]

22. Hühnel, I.; Fölster, M.; Werheid, K.; Hess, U. Empathic reactions of younger and older adults: No age related decline in affective responding. *J. Exp. Soc. Psychol.* **2014**, *50*, 136–143. [CrossRef]

23. Gunning-Dixon, F.M.; Gur, R.C.; Perkins, A.C.; Schroeder, L.; Turner, T.; Turetsky, B.I.; Chan, R.M.; Loughead, J.W.; Alsop, D.C.; Maldjian, J.; et al. Age-related differences in brain activation during emotional face processing. *Neurobiol. Aging* **2003**, *24*, 285–295. [CrossRef]

24. Tessitore, A.; Hariri, A.R.; Fera, F.; Smith, W.G.; Das, S.; Weinberger, D.R.; Mattay, V.S. Functional changes in the activity of brain regions underlying emotion processing in the elderly. *Psychiatry Res. Neuroimaging* **2005**, *139*, 9–18. [CrossRef] [PubMed]

25. Fischer, H.; Nyberg, L.; Bäckman, L. Age-related differences in brain regions supporting successful encoding of emotional faces. *Cortex* **2010**, *46*, 490–497. [CrossRef]

26. Fischer, H.; Sandblom, J.; Gavazzeni, J.; Fransson, P.; Wright, C.I.; Bäckman, L. Age-differential patterns of brain activation during perception of angry faces. *Neurosci. Lett.* **2005**, *386*, 99–104. [CrossRef] [PubMed]

27. West, J.T.; Horning, S.M.; Klebe, K.J.; Foster, S.M.; Cornwell, R.E.; Perrett, D.; Burt, D.M.; Davis, H.P. Age effects on emotion recognition in facial displays: From 20 to 89 years of age. *Exp. Aging Res.* **2012**, *38*, 146–168. [CrossRef] [PubMed]

28. Mienaltowski, A.; Groh, B.; Secula, D.; Rinne, A.; Rogers, C. Impact of expressive intensity on age differences in fear and anger detection in the periphery. *J. Vis.* **2018**, *18*, 568. [CrossRef]

29. Eimer, M.; Holmes, A. Event-related brain potential correlates of emotional face processing. *Neuropsychologia* **2007**, *45*, 15–31. [CrossRef]

30. Eimer, M.; Holmes, A.; McGlone, F.P. The role of spatial attention in the processing of facial expression: An ERP study of rapid brain responses to six basic emotions. *Cogn. Affect. Behav. Neurosci.* **2003**, *3*, 97–110. [CrossRef]

31. Kisley, M.A.; Wood, S.; Burrows, C.L. Looking at the sunny side of life: Age-related change in an event-related potential measure of the negativity bias. *Psychol. Sci.* **2007**, *18*, 838–843. [CrossRef]

32. Wieser, M.J.; Mühlberger, A.; Kenntner-Mabiala, R.; Pauli, P. Is emotion processing affected by advancing age? An event-related brain potential study. *Brain Res.* **2006**, *1096*, 138–147. [CrossRef] [PubMed]

33. Tsolaki, A.C.; Kosmidou, V.E.; Kompatsiaris, I.Y.; Papadaniil, C.; Hadjileontiadis, L.; Tsolaki, M. Age-induced differences in brain neural activation elicited by visual emotional stimuli: A high-density EEG study. *Neuroscience* **2017**, *340*, 268–278. [CrossRef] [PubMed]

34. Nelson, D.B. Conditional heteroskedasticity in asset returns: A new approach. *Econom. J. Econom. Soc.* **1991**, *59*, 347–370. [CrossRef]

35. Turner, M.; Ridsdale, J. The digit memory test. *Dyslexia Int.* **2004**.

36. Bagby, R.M.; Parker, J.D.; Taylor, G.J. The twenty-item Toronto Alexithymia Scale—I. Item selection and cross-validation of the factor structure. *J. Psychosom. Res.* **1994**, *38*, 23–32. [CrossRef]

37. Reuter-Lorenz, P.A.; Cappell, K.A. Neurocognitive aging and the compensation hypothesis. *Curr. Dir. Psychol. Sci.* **2008**, *17*, 177–182. [CrossRef]

38. Folstein, M.F.; Folstein, S.E.; McHugh, P.R. "Mini-mental state": A practical method for grading the cognitive state of patients for the clinician. *J. Psychiatr. Res.* **1975**, *12*, 189–198. [CrossRef]

39. Chayer, C. The neurologic examination: Brief mental status. *Can. J. Geriatr. Care* **2002**, *1*, 265–267.

40. Roesch, E.B.; Tamarit, L.; Reveret, L.; Grandjean, D.; Sander, D.; Scherer, K.R. FACSGen: A tool to synthesize emotional facial expressions through systematic manipulation of facial action units. *J. Nonverbal Behav.* **2011**, *35*, 1–16. [CrossRef]

41. Oosterhof, N.N.; Todorov, A. Shared perceptual basis of emotional expression and trustworthiness impression from faces. *Emotion* **2009**, *9*, 128–133. [CrossRef]

42. Penton, T.; Bate, S.; Dalrymple, K.; Reed, T.; Kelly, M.; Godovich, S.; Tamm, M.; Duchaine, B.; Banissy, M.J. Using high frequency transcranial random noise stimulation to modulate face memory performance in younger and older adults: Lessons learnt from mixed findings. *Front. Neurosci.* **2018**, *12*, 863. [CrossRef] [PubMed]

43. Delorme, A.; Makeig, S. EEGLAB: An open source toolbox for analysis of single-trial EEG dynamics including independent component analysis. *J. Neurosci. Methods* **2004**, *134*, 9–21. [CrossRef] [PubMed]

44. Oostenveld, R.; Fries, P.; Maris, E.; Schoffelen, J.M. FieldTrip: Open source software for advanced analysis of MEG, EEG, and invasive electrophysiological data. *Comput. Intell. Neurosci.* **2011**, *1*, 1–9. [CrossRef] [PubMed]

45. Picton, T.W.; van Roon, P.; Armilio, M.L.; Berg, P.; Ille, N.; Scherg, M. Blinks, saccades, extraocular muscles and visual evoked potentials (Reply to Verleger). *J. Psychophysiol.* **2000**, *14*, 210–217. [CrossRef]

46. Pfurtscheller, G.; Da Silva, F.L. Event-related EEG/MEG synchronization and desynchronization: Basic principles. *Clin. Neurophysiol.* **1999**, *110*, 1842–1857. [CrossRef]

47. Maris, E.; Oostenveld, R. Nonparametric statistical testing of EEG-and MEG-data. *J. Neurosci. Methods* **2007**, *164*, 177–190. [CrossRef]

48. Lindsen, J.P.; Jones, R.; Shimojo, S.; Bhattacharya, J. Neural components underlying subjective preferential decision making. *Neuroimage* **2010**, *50*, 1626–1632. [CrossRef]

49. Luft, C.D.B.; Takase, E.; Bhattacharya, J. Processing graded feedback: Electrophysiological correlates of learning from small and large errors. *J. Cogn. Neurosci.* **2014**, *26*, 1180–1193. [CrossRef]

50. Sandkühler, S.; Bhattacharya, J. Deconstructing insight: EEG correlates of insightful problem solving. *PLoS ONE* **2008**, *3*, 1459. [CrossRef]

51. MacPherson, S.E.; Phillips, L.H.; Sala, S.D. Age-related differences in the ability to perceive sad facial expressions. *Aging Clin. Exp. Res.* **2006**, *18*, 418–424. [CrossRef]

52. Gratton, G.; Coles, M.G.; Donchin, E. Optimizing the use of information: Strategic control of activation of responses. *J. Exp. Psychol. Gen.* **1992**, *121*, 480. [CrossRef] [PubMed]

53. Schmidt, J.R. Questioning conflict adaptation: Proportion congruent and Gratton effects reconsidered. *Psychon. Bull. Rev.* **2013**, *20*, 615–630. [CrossRef] [PubMed]

54. Aschenbrenner, A.J.; Balota, D.A. Dynamic adjustments of attentional control in healthy aging. *Psychol. Aging* **2017**, *32*, 1–15. [CrossRef] [PubMed]

55. Lamont, A.C.; Stewart-Williams, S.; Podd, J. Face recognition and aging: Effects of target age and memory load. *Mem. Cogn.* **2005**, *33*, 1017–1024. [CrossRef]

56. Eimer, M.; Holmes, A. An ERP study on the time course of emotional face processing. *Neuroreport* **2002**, *13*, 427–431. [CrossRef]

57. Balconi, M.; Pozzoli, U. Face-selective processing and the effect of pleasant and unpleasant emotional expressions on ERP correlates. *Int. J. Psychophysiol.* **2003**, *49*, 67–74. [CrossRef]

58. Raz, N. Aging of the brain and its impact on cognitive performance: Integration of structural and functional findings. In *The Handbook of Aging and Cognition*; Craik, F.I.M., Salthouse, T.A., Eds.; Lawrence Erlbaum Associates Publishers: Mahwah, NJ, USA, 2000; pp. 1–90.

59. Grady, C.L. Functional brain imaging and age-related changes in cognition. *Biol. Psychol.* **2000**, *54*, 259–281. [CrossRef]
60. Halpern, A.R.; Zioga, I.; Shankleman, M.; Lindsen, J.; Pearce, M.T.; Bhattacharya, J. That note sounds wrong! Age-related effects in processing of musical expectation. *Brain Cogn.* **2017**, *113*, 1–9. [CrossRef]
61. Johannsen, P.; Jakobsen, J.; Bruhn, P.; Hansen, S.B.; Gee, A.; Stødkilde-Jørgensen, H.; Gjedde, A. Cortical sites of sustained and divided attention in normal elderly humans. *Neuroimage* **1997**, *6*, 145–155. [CrossRef]
62. Madden, D.J.; Turkington, T.G.; Provenzale, J.M.; Hawk, T.C.; Hoffman, J.M.; Coleman, R.E. Selective and divided visual attention: Age-related changes in regional cerebral blood flow measured by H2 15O PET. *Hum. Brain Mapp.* **1997**, *5*, 389–409. [CrossRef]
63. Anderson, N.D.; Iidaka, T.; Cabeza, R.; Kapur, S.; McIntosh, A.R.; Craik, F.I. The effects of divided attention on encoding-and retrieval-related brain activity: A PET study of younger and older adults. *J. Cogn. Neurosci.* **2000**, *12*, 775–792. [CrossRef] [PubMed]
64. Hartley, A.A.; Speer, N.K.; Jonides, J.; Reuter-Lorenz, P.A.; Smith, E.E. Is the dissociability of working memory systems for name identity, visual-object identity, and spatial location maintained in old age? *Neuropsychology* **2001**, *15*, 3. [CrossRef] [PubMed]
65. Mitchell, K.J.; Johnson, M.K.; Raye, C.L.; Mather, M.; D'esposito, M. Aging and reflective processes of working memory: Binding and test load deficits. *Psychol. Aging* **2000**, *15*, 527. [CrossRef] [PubMed]
66. Smith, E.E.; Geva, A.; Jonides, J.; Miller, A.; Reuter-Lorenz, P.; Koeppe, R.A. The neural basis of task-switching in working memory: Effects of performance and aging. *Proc. Natl. Acad. Sci. USA* **2001**, *98*, 2095–2100. [CrossRef] [PubMed]
67. Gazzaley, A.; Cooney, J.W.; Rissman, J.; D'Esposito, M. Top-down suppression deficit underlies working memory impairment in normal aging. *Nat. Neurosci.* **2005**, *8*, 1298–1300. [CrossRef] [PubMed]
68. Phelps, E.A.; Ling, S.; Carrasco, M. Emotion facilitates perception and potentiates the perceptual benefits of attention. *Psychol. Sci.* **2006**, *17*, 292–299. [CrossRef]
69. Houston, J.R.; Pollock, J.W.; Lien, M.C.; Allen, P.A. Emotional arousal deficit or emotional regulation bias? An electrophysiological study of age-related differences in emotion perception. *Exp. Aging Res.* **2018**, *44*, 187–205. [CrossRef]
70. Schupp, H.T.; Cuthbert, B.N.; Bradley, M.M.; Cacioppo, J.T.; Ito, T.; Lang, P.J. Affective picture processing: The late positive potential is modulated by motivational relevance. *Psychophysiology* **2000**, *37*, 257–261. [CrossRef]
71. Van der Gaag, C.; Minderaa, R.B.; Keysers, C. Facial expressions: What the mirror neuron system can and cannot tell us. *Soc. Neurosci.* **2007**, *2*, 179–222. [CrossRef]
72. Jabbi, M.; Keysers, C. Inferior frontal gyrus activity triggers anterior insula response to emotional facial expressions. *Emotion* **2008**, *8*, 775. [CrossRef]
73. Kriegeskorte, N.; Simmons, W.K.; Bellgowan, P.S.; Baker, C.I. Circular analysis in systems neuroscience: The dangers of double dipping. *Nat. Neurosci.* **2009**, *12*, 535. [CrossRef] [PubMed]
74. Fusar-Poli, P.; Placentino, A.; Carletti, F.; Landi, P.; Allen, P.; Surguladze, S.; Benedetti, F.; Abbamonte, M.; Gasparotti, R.; Barale, F.; et al. Functional atlas of emotional faces processing: A voxel-based meta-analysis of 105 functional magnetic resonance imaging studies. *J. Psychiatry Neurosci.* **2009**, *34*, 418–432. [PubMed]
75. Kesler, M.L.; Andersen, A.H.; Smith, C.D.; Avison, M.J.; Davis, C.E.; Kryscio, R.J.; Blonder, L.X. Neural substrates of facial emotion processing using fMRI. *Cogn. Brain Res.* **2001**, *11*, 213–226. [CrossRef]

Article

Using an Overlapping Time Interval Strategy to Study Diagnostic Instability in Mild Cognitive Impairment Subtypes

David Facal [1,*], Joan Guàrdia-Olmos [2], Arturo X. Pereiro [1], Cristina Lojo-Seoane [1], Maribel Peró [2] and Onésimo Juncos-Rabadán [1]

[1] Department of Developmental Psychology, University de Santiago de Compostela, 15782 Santiago de Compostela, Galicia, Spain; arturoxose.pereiro@usc.es (A.X.P.); cristina.lojo@usc.es (C.L.-S.); onesimo.juncos@usc.es (O.J.-R.)

[2] Department of Methodology of Behavioural Sciences, University of Barcelona, 08035 Barcelona, Catalunya, Spain; jguardia@ub.edu (J.G.-O.); mpero@ub.edu (M.P.)

* Correspondence: david.facal@usc.es

Received: 21 August 2019; Accepted: 18 September 2019; Published: 19 September 2019

Abstract: (1) Background: Mild cognitive impairment (MCI) is a diagnostic label in which stability is typically low. The aim of this study was to examine temporal changes in the diagnosis of MCI subtypes by using an overlapping-time strategy; (2) Methods: The study included 435 participants aged over 50 years with subjective cognitive complaints and who completed at least one follow-up evaluation. The probability of transition was estimated using Bayesian odds ratios; (3) Results: Within the different time intervals, the controls with subjective cognitive complaints represented the largest proportion of participants, followed by sda-MCI at baseline and in the first five intervals of the follow-up, but not in the last eight intervals. The odds ratios indicated higher odds of conversion to dementia in sda-MCI and mda-MCI groups relative to na-MCI (e.g., interval 9–15 months—sda-MCI OR = 9 and mda-MCI OR = 3.36; interval 27–33—sda-MCI OR = 16 and mda-MCI OR = 5.06; interval 42–48—sda-MCI OR = 8.16 and mda-MCI = 3.45; interval 45–51—sda-MCI OR = 3.31 and mda-MCI = 1); (4) Conclusions: Notable patterns of instability consistent with the current literature were observed. The limitations of a prospective approach in the study of MCI transitions are discussed.

Keywords: cognitive aging; mild cognitive impairment; subjective cognitive complaints; conversion to dementia; Bayesian odds ratios; time overlapping intervals; screening and diagnosis

---

1. Introduction

Mild cognitive impairment (MCI) is the diagnostic entity used to describe a condition in middle-aged and old adults who experience cognitive decline but without impaired daily functioning [1]. MCI has been indicated as a focal concept for understanding pre-dementia stages in relation to cognitive ageing [2]. Despite some controversy, the following are generally accepted as the core MCI criteria: (i) Evidence of subjective cognitive complaints from patients or their close contacts; (ii) evidence of cognitive impairment in one or more cognitive domains that is greater than expected for the patient's age and educational background; (iii) preservation or minimal impairment of instrumental activities of daily living; and (iv) non-fulfillment of diagnostic criteria for dementia [3–7]. This consensus also extends to the different subtypes of MCI (amnestic and non-amnestic, single and multi-domain) proposed by Petersen and colleagues [3,6]. Apart from its importance in research on cognitive impairment, MCI is also a multifaceted clinical label, characterized by complex cognitive changes and diagnostic instability in MCI subtypes, conversion to dementia and recovery to normal cognitive aging [8–10].

Although major advances in research have led to corresponding changes in the diagnostic criteria for MCI, it remains unclear how cognitive deficits increase over time and how these deficits affect progression to dementia in different MCI subtypes [1,11]. According to Gersteneker and Mast [1], the relationship between MCI subtypes and transition and conversion patterns is not as linear as initially proposed. Amnestic MCI is considered to be the MCI subtype most likely to progress to AD. The diagnostic guidelines produced by the National Institute on Aging- Alzheimer's disease working groups [5] indicate that the 'MCI, due to AD', refers to the symptomatic pre-dementia phase of AD and is characterized by impairment in one or more cognitive domains, typically including episodic memory. Different studies have already shown that multi-domain amnestic MCI (mda-MCI) is the most reliable MCI subtype, with a lower chance of reversion and a higher risk of conversion to dementia over time [12,13].

Although MCI is understood to be an unstable stage and the passage of the time is relevant in its cognitive and clinical manifestations, scant attention has been given in the literature to the time between assessments -but see [9,14]. Establishing this interval may have a considerable impact on the findings of MCI research. For example, longitudinal studies have developed non-linear, plateau models of decline that are dependent on the time of measurement [15]. In the current literature on MCI, the date of initial diagnosis and subsequent measurement points are determined as a function of the study start date; this is problematic as the moment at which participants report subjective cognitive complaints (SCCs) is intrinsically variable [6]. To address this problem, Cloutier et al. [11] suggested aligning the moment retrospectively according to the year in which participants received their diagnosis of AD. Facal et al. [9] proposed an alternative, complementary approach involving an overlapping interval strategy that considers different mid-point stages depending on the time between the baseline and follow-up assessment.

The aim of the present study was to test the capacity of an overlapping time interval strategy to address changes in MCI, by including supplementary follow-up evaluations and broader time intervals. The overlapping interval strategy was applied in the current sample of each time interval to explore whether changes (decline, stability, recovery) are temporally stable or vary according to different time intervals, and also to examine the trajectories of cognitive function in different MCI subtypes within different intervals.

2. Materials and Methods

2.1. Participants

The current research included 435 participants over 50 years old (range 50–88) included in the on-going Compostela aging study and who completed baseline and at least one follow-up evaluation [16]. All participants were referred to us by general practitioners after attending primary care health centers with subjective cognitive complaints, but with no prior diagnosis of dementia, psychiatric or neurological disorders.

At baseline, the sample comprised 40 participants in the mda-MCI group (multi-domain amnestic MCI), 34 participants in the na-MCI group (no amnestic MCI), 68 participants in the sda-MCI, and 293 participants in the control group with subjective cognitive complaints (SCCs). Age, years of education and results of the cognitive assessment at baseline are shown in Table 1. Group differences were calculated using non-parametric tests (Kruskal-Wallis and Mann-Whitney test) given the differences in sample size between the groups. Diagnoses were made following the Petersen criteria [3,4] updated by Albert et al. [5] and were reached by consensus at research team diagnostic meetings.

Table 1. Mean values and standard deviations (in parentheses) of the demographic and cognitive measures in each group at baseline.

	mda-MCI	na-MCI	sda-MCI	SCCs	Group Differences χ^2 (gl)	Group Comparisons
Age	72.42 (8.33)	67.12 (8.82)	69.29 (9.36)	65.49 (9.05)	24.83 **	mda-MCI > naMCI, SCCs; sda-MCI>SCCs
Years of education	9.68 (8.82)	8.29 (3.75)	9.40 (4.17)	9.92 (4.70)	1.81	
Memory complaints—participant	19.43 (4.69)	20.35 (3.45)	19.03 (4.69)	18.81 (4.51)	8.53 *	naMCI > SCCs
Memory complaints—proxy	17.97 (4.54)	18.03 (4.76)	16.84 (4.46)	15.49 (4.23)	17.03 **	mda-MCI, naMCI > sdaMCI, SCCs; sda-MCI > SCCs
CVLT Short Delay Free Recall	3.10 (2.03)	9.18 (2.21)	3.88 (2.05)	10.34 (2.70)	228.58 **	SCCs, na-MCI > mda-MCI, sda-MCI; sda-MCI > mda-MCI
CVLT Long Delay Free Recall	3.82 (3.19)	9.76 (2.69)	5.12 (3.06)	11.14 (2.83)	184.80 **	SCCs, sda-MCI, na-MCI > mdaMCI; SCCs> na-MCI, sda-MCI; na-MCI>sda-MCI
CAMCOG-R memory	15.50 (3.80)	18.56 (2.87)	18.53 (3.88)	21.26 (2.78)	99.18 **	SCCs, sda-MCI, na-MCI > mdaMCI; SCCs > na-MCI, sda-MCI
CAMCOG-R orientation	7.98 (1.46)	9.29 (0.80)	9.31 (0.83)	9.64 (0.63)	79.19 **	SCCs, sda-MCI, na-MCI > mdaMCI; SCCs > na-MCI, sda-MCI
CAMCOG-R language	22.77 (2.36)	23.53 (2.00)	24.90 (2.43)	25.60 (2.40)	54.55 **	SCCs, sda-MCI > mdaMCI, na-MCI
CAMCOG-R attention - calculation	5.03 (2.21)	4.29 (1.73)	7.25 (1.70)	7.05 (1.97)	100.90 **	SCCs, sda-MCI > mdaMCI, na-MCI
CAMCOG-R praxis	9.10 (2.53)	10.03 (1.71)	10.71 (1.47)	11.14 (1.19)	48.01 **	SCCs, sda-MCI > mdaMCI, na-MCI
CAMCOG-R perception	5.97 (1.64)	6.50 (1.28)	6.53 (1.48)	7.06 (1.44)	21.68 **	SCCs, sda-MCI > mdaMCI; SCCs > na-MCI, sda-MCI
CAMCOG-R executive function	13.22 (3.94)	15.35 (4.20)	15.93 (4.28)	18.42 (4.20)	58.84 **	SCCs, sda-MCI > mdaMCI; SCCs > na-MCI, sda-MCI

** $p < 0.01$ * $p < 0.05$. Note: CVLT = california verbal learning test; CAMCOG-R = cambridge cognitive assessment-revised; mda-MCI = multi-domain amnestic Mild Cognitive Impairment; na-MCI = non-amnestic mild cognitive impairment; sda-MCI = single-domain amnestic mild cognitive impairment; SCCs = subjective cognitive complaints.

In the final sample that completed at least one follow up evaluation, 415 participants completed the first follow-up around $1_{1/2}$ year after baseline (mean = 17.92 months, S.D. = 3.87, range 10–31 months), and 330 participants completed the second follow up assessment around 3 years after baseline (mean = 37.13 months, S.D. = 5.43, range 27–51 months; this includes 20 participants who did not complete the first follow-up assessment, but who did participate in the second follow-up). Reasons for deviation from the reference times include holidays, health issues and work issues.

2.2. Procedure

The baseline evaluations were made between 2 January 2008 and 11 November 2012. Participants undertook wide-ranging cognitive and neuropsychological evaluations, including the Spanish versions of the California Verbal Learning Test (CVLT) [17], the Mini-Mental State Examination (MMSE) [18] and the Cambridge Cognitive Examination—Revised (CAMCOG-R), which include subscales in several cognitive domains (memory, orientation, language, attention and calculation, praxis, perception and executive functioning) [19]. Within the follow-up assessments, all participants again underwent the assessment and were re-diagnosed by the same research team, again by consensus at specific meetings.

At all three evaluation times, the cut-off was 1.5 standard deviations (SDs) below age and education norms on the corresponding tests. All participants who displayed normal cognitive performance (scores higher than cut-off scores on cognitive status and cognitive functioning, including memory) were included in the SCCs group, as they all reported subjective memory complaints. For all MCI participants, the general criteria outlined by Albert et al. [5] were applied: (a) Presence of complaints corroborated by an informant; (b) impairment in one or more cognitive functions; (c) independence in daily living with minimum support or help; and (d) absence of dementia according to the NINCDS-ADRDA and DSM-IV standards. For the specific diagnosis of MCI subtypes, the criteria proposed by Petersen and colleagues [3,6] were applied, including (b1) a score of 1.5 SDs below age standards for short-term and long-term recall measures of the Spanish version of the CVLT for amnestic MCI (aMCI). Two different types of aMCI were diagnosed: Sda-MCI in those participants (b2) who performed 1.5 SDs below age norms in the Spanish version of the CVLT, but performed normally in the MMSE and the CAMCOG-R subscales; and mda-MCI in those participants (b3) who performed 1.5 SDs below norms in the MMSE and at least two sub-scores of the CAMCOG-R representing at least two cognitive functions. Finally, na-MCI was diagnosed in those participants (b4) who performed within the normal range in the Spanish version of the CVLT, but 1.5 SDs below average in at least one of the other subscales of the CAMCOG-R. Single- and multi-domain na-MCI were not differentiated, because of the small sample size in the single-domain na-MCI group. Probable AD or other types of dementia were diagnosed according to the NINCDS-ADRDA and DMS-IV criteria. Progression to dementia was confirmed by consultation of the medical history, and the date of neurological diagnosis was recorded.

The study received approval from the Ethics in Clinical Research Committee of the Galician Government and was conducted in accordance with the provisions of the Declaration of Helsinki as revised in Brazil 2013. Written informed consent was obtained from all participants.

2.3. Overlapping Intervals Procedure

According to the variability in the time between the baseline and follow-up evaluations, and the potential effect of time in the study of MCI according to its variability, a time-overlapping interval approach was taken [20] considering 13 mid-point time intervals with a range of 6 months across each interval. This approach enabled a wide range of time distributions to be covered between assessments and also consideration of all participants evaluated, irrespective of the time at which the follow-up was carried out and whether they participated in the first follow up assessment (about one year and a half after the baseline assessment) and/or the second follow up (about three years after the baseline). Seven intervals were included within the time-lapse established for the first follow up assessment (9–15, 12–18, 15–21, 18–24, 21–27, 24–30, 27–33), and seven intervals were also included within the time-lapse established for the second follow up assessment (27–33, 30–36, 33–39, 36–42, 39–45, 42–48, 45–51).

According to this overlapping-interval design, and in order to maximize sample sizes and temporal distributions across successive evaluations, the sample size was different for each interval (see Table 2). According to the reference times for the first (18 months) and the second (36 months) follow-up assessments, the sample sizes tended to be larger in the intervals closer to these times and smaller in the intervening intervals. This approach allows overlapping counts between time intervals, avoiding double-counting of participants by considering the participants only once in those intervals in which months between assessments are included. The range around each mid-point interval was six months, so that adjacent intervals differ in mean time between assessments, but overlap in the time range. Therefore, the mid-point of the 9–15 month group was 12, the mid-point for the 12–18 month group was 15, and so on. The results can, thus, be consulted with the maximum temporal continuum (13 time groups) or with the minimum overlap (seven and six time groups, 9–15, 15–21, 21–27, 27–33, 33–39, 39–45, 45–51 months and 12–18, 18–24, 24–30, 30–36, 36–42, 42–48, respectively). Overlapping intervals are used as independent sample estimates.

Table 2. Sample size in each overlapped time interval.

Interval	mda-MCI	na-MCI	sda-MCI	SCCs
9–15 months	6	3	10	35
12–18 months	14	15	30	125
15–21 months	16	16	31	155
18–24 months	13	11	22	105
21–27 months	5	8	13	49
24–30 months	4	4	7	25
27–33 months	6	3	11	47
30–36 months	9	7	21	98
33–39 months	10	15	21	98
36–42 months	8	12	14	56
39–45 months	8	8	9	51
42–48 months	6	3	9	27
45–51 months	4	4	9	18

2.4. Statistical Analysis

The probability of transition from one state to another was estimated using Bayesian estimates, calculated as follows:

$$P(A_i|B) = p(B|A_i)P(A_i)/\Sigma_{k=1} P(B|A_k)P(A_k),$$

where P(Ai) represents the a priori probability, P(B|Ai) represents the probability of transition from Ai to B and, finally, P(Ai|B) represents the a-posteriori probability. In the present study, event Ai assumes the state of the diagnosis at a given time (*t*), and B is the state of diagnosis at time (*t* +1). Evidently, transitions from states Ai to Ak and then to the next state B draw all combinations existing between the diagnostic categories at a given moment and the same categories at the next time. Models of these transitions were generated using P(Ai) and P(Ai|B) from estimates of the proportions observed in the original distributions.

In addition, as the demands of the Bayes' theorem require exhaustiveness, the odds ratio was obtained from the contrasts, as follows:

$$ORij = P(Ai|B)/P(Aj|B),$$

with the aim of evaluating the most probable transitions of the pair (i,j) as in the previous expression. Thus, apart from the estimated values of the Bayesian probabilities, more applied values of higher clinical value were obtained. Details of all of the calculations, including those made to obtain Odds Ratios in each interval, are available to readers in Supplementary File S1.

All the mathematical procedures were generated by Excel routines and specific programming in MatLab.

3. Results

The diagnostic probabilities for the different MCI subtypes are presented in Figures 1 and 2. Figure 3 shows those participants whose diagnosis did not change in the follow-up assessments and those participants in the different groups who progressed to dementia. The number of participants in the mda-MCI group who progressed to dementia was 4 at the first follow-up assessment and 7 at the second follow-up assessment; the corresponding numbers in the na-MCI group were 0 at the first follow-up and 1 at the second follow-up assessment; the numbers in the sda-MCI group were 3 at the first follow-up and 5 at the second follow-up; and finally, the corresponding numbers in the SCCs group were 0 at the first follow-up and 4 at the second follow-up. Cases recorded at the first follow-up were also counted at the second follow-up, so that the numbers are cumulative.

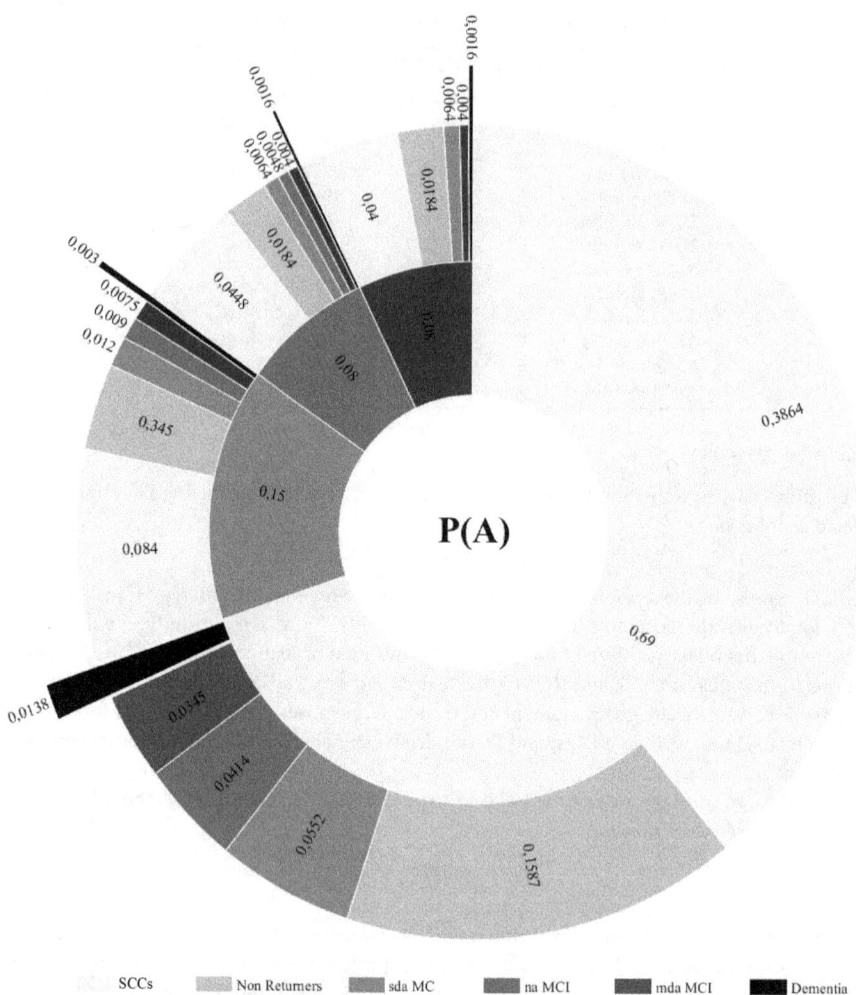

Figure 1. Diagnostic probabilities at baseline (inner circle) and at first follow-up (outer circle). Diagnostic probabilities are presented numerically, as proportions, with the total sum of proportions in the inner and outer circles equaling 1. The naMCI group is absent from the outer circle, but corresponds to the mdaMCI group in the inner circle as no transitions from mdaMCI to naMCI were recorded. SCCs = Subjective cognitive complaints; MCI = mild cognitive impairment; mda = multidomain amnestic; na = non-amnestic; sda = single domain amnestic.

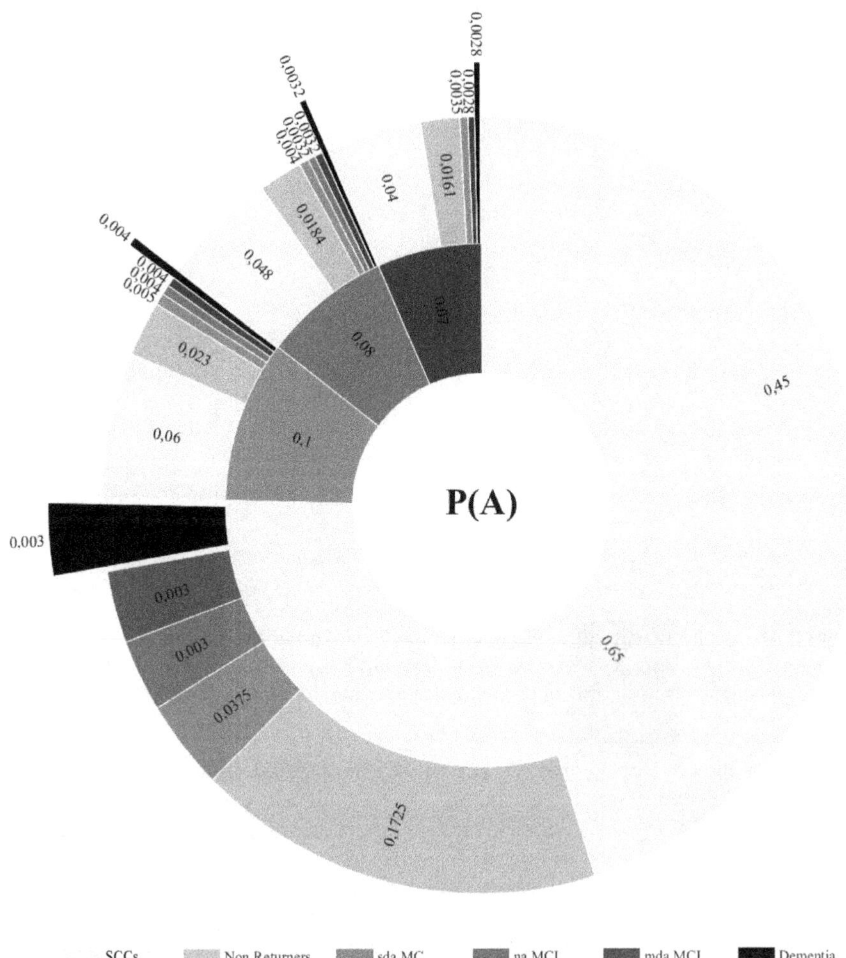

Figure 2. Diagnostic probabilities at first follow-up (inner circle) and at second follow-up (outer circle). Diagnostic probabilities are presented numerically, as proportions, with the total sum of proportions in the inner and outer circles equaling 1. The naMCI group is absent from the outer circle, but corresponds to the mdaMCI group in the inner circle as no transitions from mdaMCI no naMCI were recorded. SCCs = Subjective cognitive complaints; MCI = mild cognitive impairment; mda = multi-domain amnestic; na = non-amnestic; sda = single domain amnestic.

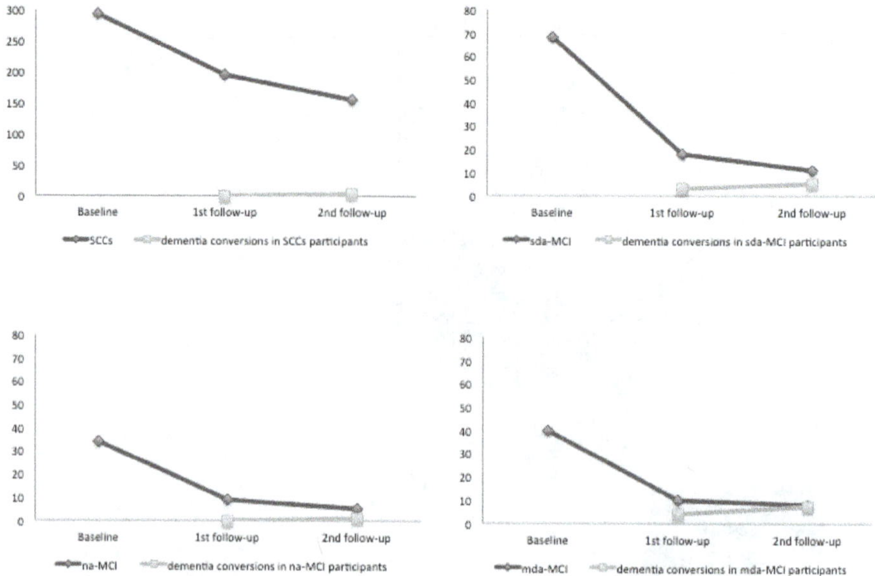

Figure 3. The number of participants whose diagnosis did not change at the follow-up assessments, and the number of participants who converted to dementia. SCCs = Subjective cognitive complaints; MCI = mild cognitive impairment; mda = multi-domain amnestic; na = non-amnestic; sda = single domain amnestic.

According to the overlapping time interval strategy, the diagnostic probabilities for the different MCI subtypes at different times are presented in Figure 4. Similar trends were observed for the different time intervals, as follows: (a) SCCs represented the largest proportion of participants, and the proportion tended to increase at follow-up assessments; (b) sda-MCI participants comprised the second largest group at baseline and in the first five intervals of the follow-up, but not in the last eight intervals of the follow-up continuum; (c) the probabilities of diagnosis of mda-MCI and na-MCI were similar at both baseline and follow-up, with slight variations across the time intervals; and (d) there was a small, but increasing, probability of conversion to dementia, represented in the right column of the follow-up section in Figure 3. In this figure, only the cases of conversion to dementia recorded during each time interval are counted.

According to the higher, but temporally variable, probability of sda-MCI at follow-up and to test the theoretical relevance of mda-MCI as the closest point to dementia within the MCI continuum, odds ratios were calculated for each time interval and for sda-MCI and mda-MCI relative to na-MCI. This enabled the probability of conversion to dementia to be determined by comparing the two different amnestic subtypes and considering the non-amnestic subtype as the reference subtype. The conversion odds (Table 3) revealed a higher probability of conversion to dementia in sda-MCI than in na-MCI, which would be expected according to the higher proportion of sda-MCI participants at baseline, and also a higher probability of conversion to dementia in mda-MCI than in na-MCI.

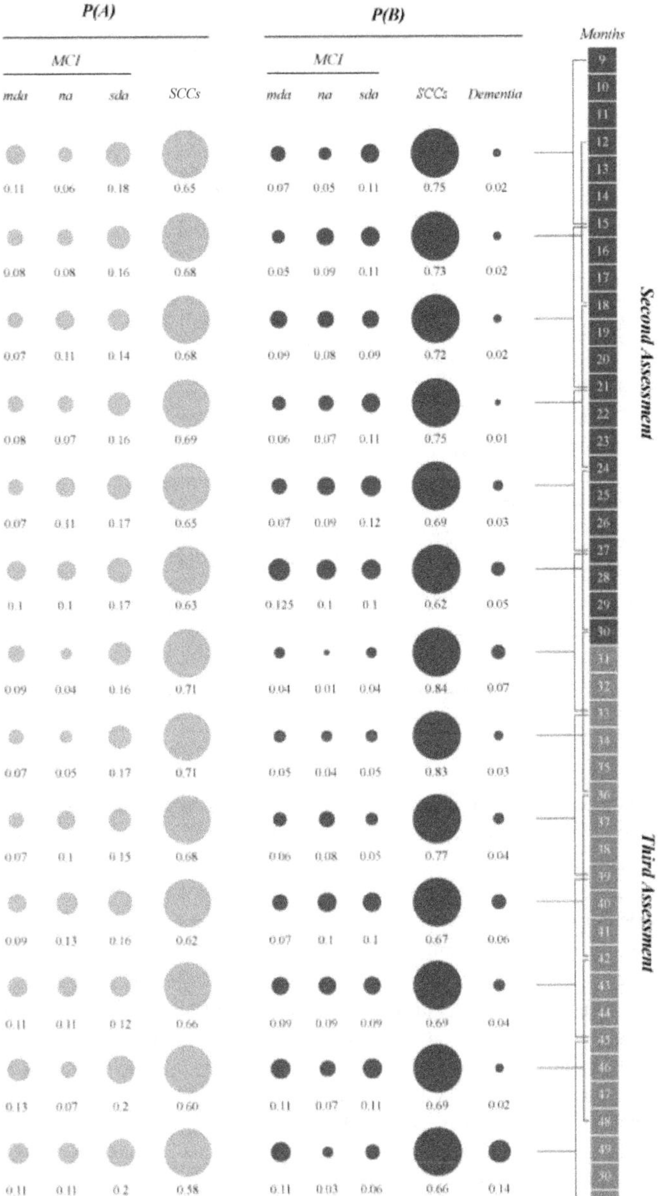

Figure 4. Diagnostic probabilities at baseline **P(A)** and at follow-up **P(B)**. Months between baseline and follow-up assessments are represented on the right-hand side of the diagram. Each line represents one time interval, with brackets in the column on the right showing the overlapping nature of the intervals (e.g., time interval 1 is indicated by the bracket encompassing months 9 to 15 and overlaps with interval 2, indicated by the bracket encompassing months 12 to 18). Diagnostic probabilities are represented at baseline in light grey and at follow-up in dark grey. SCCs = subjective cognitive complaints; MCI = mild cognitive impairment; mda = multidomain amnestic; na = non-amnestic; sda = single domain amnestic.

Table 3. Odds ratios for conversion to dementia in sda-MCI and mda-MCI groups relative to na-MCI.

Interval	sda-MCI OR (95% CI)	mda-MCI OR (95% CI)
9–15 months	9 (8.20–9.80)	3.36 (3.06–3.66)
12–18 months	4 (3.35–4.65)	1 (1.00–2.12)
15–21 months	1.62 (1.22–2.02)	0.41 (0–1.21)
18–24 months	5.22 (4.67–5.77)	1.31 (1.08–1.34)
21–27 months	2.39 (1.45–3.34)	0.41 (0–1.21)
24–30 months	2.89 (2.10–3.68)	1 (1.00–2.12)
27–33 months	16 (12.10–19.90)	5.06 (4.85–5.27)
30–36 months	11.56 (10.10–12.45)	1.96 (1.52–2.40)
33–39 months	2.25 (1,10–3.40)	0.49 (0–1.21)
36–42 months	1.51 (1.10–1.92)	0.48 (0–1.21)
39–45 months	1.19 (1.00–1.77)	1 (1.00–2.12)
42–48 months	8.16 (6.80–9.54)	3.45 (2.90–4.05)
45–51 months	3.31 (2.80–3.82)	1 (1.00–2.12)

4. Discussion

This study overcomes some limitations of previous research considering diagnostic transition within MCI subtypes and conversion to dementia within different time frames. For example, it includes longer follow-up periods than in previous studies e.g., [9,13] and it uses Bayesian estimates to illustrate the diagnostic transitions more clearly and accurately, representing the reality of daily practice by avoiding an artificial balance of the probability of diagnosis at baseline. In daily gerontological and geriatric practice, longitudinal follow-up and serial cognitive assessments are recommended as they provide a clearer picture of the patient's baseline and trajectory of cognitive function over time [21], although the optimal timing and cost-effectiveness of longitudinal cognitive assessments remain unclear. In this regard, the transition probabilities calculated using the Bayesian approach with overlapping-interval design are similar to current daily practice in which the time of assessment is variable (i.e., the higher odds in sda-MCI than in mda-MCI compared to na-MCI can be explained by the higher proportion of sda-MCI participants at baseline, in parallel to the higher number of patients with sda-MCI that professionals diagnose in daily practice).

The increase in number of participants classified as SCCs and the decrease in those diagnosed as sda-MCI in the second part of the follow-up assessments indicates the instability of this MCI subtype, as previously highlighted by Malek-Ahmadi in a meta-analysis [22], in which the rate of reversion to normal cognition was approximately 24%. The increase may indicate an excess number of false-positives in diagnosing this particular MCI subtype, which includes patients with impairments only in memory. Nonetheless, the higher odds ratios for conversion to dementia of sda-MCI relative to na-MCI also indicate an important risk of progression of cognitive impairment in the sda-MCI subtype [5,7]. The current clinical criteria, which include a cut-off of <1.5 SD below mean only in memory scores, indicate an unstable subtype, especially when the follow-up assessments are carried out after short intervals. This confirms that other functional, behavioral and biological markers should be included for more accurate diagnosis, as recommended by the American Academy of Neurology (AAN) guidelines on MCI [7]. Follow-up studies are needed because studies that test the actual cognitive status of middle-aged and older adults living in the community with subjective memory complaints appear to be susceptible to false-positive diagnostic errors, mainly regarding single domain MCI subtypes. In this respect, Han et al. [10] recommended considering the diagnostic stability over time and the multiplicity of impaired cognitive domains for managing MCI.

Regarding the rates of conversion to dementia, longitudinal studies have found that non-linear, plateau models of decline (initial decline in memory, followed by a temporary plateau occurring prior to a final rapid decline immediately prior to AD diagnosis) are more accurate than linear models, at least for memory performance [15]. The small, but increasing, probability of conversion to dementia between time intervals observed in our study does not support these findings, although the intervals are relatively short and a slight increase in the case of cognitive decline was detected in the final intervals. However, the rates of conversion to dementia in this study are relatively low [9,23]. Accordingly, adjusting the overlapping time interval strategy included here by including longer intervals and successive follow-up assessments may help to clarify these trends in conversion, e.g., [24].

As already mentioned, this study has some limitations, due to the temporal approach and the rate of variability in the diagnosis of MCI subtypes. Regarding the first limitation, this study used a prospective approach to the complex phenomena of MCI transition and conversion to dementia, adopting overlapping intervals to study the role of time in the MCI diagnostic process. Although the prospective approach has traditionally been used in this field, retrospective approaches involving analysis of data from the moment when patients are diagnosed with Alzheimer's disease or other types of dementia are increasingly common [6]. For the diagnosis of MCI subtypes, Klekociuk and Summers [13] indicate that it may be necessary to wait until single-domain variants transition to multiple-domain variants in order to identify genuine cases, particularly when recruiting prospective samples with subjective memory complaints—as in the present study. Regarding the variability in diagnosis of MCI subtypes, in this study, we measured verbal learning and memory with the Spanish version of the CVLT [9] and other cognitive functions with the CAMCOG-R. The use of other specific cognitive measures of memory and non-memory functions [25,26], as well as functional, behavioral and biological measures, could minimize the diagnostic instability [7]. It has been suggested that the number of false-positive can also be reduced by using empirical statistical approaches to identify MCI, such as cluster analysis based on the actual cognitive performance of the participants rather than previously established cut-off scores [27–29]. In this regard, Edmonds et al. [27] suggest that the inclusion of self-reported cognitive complaints as a core MCI diagnostic criterion may yield higher false-positive rates, as amnestic MCI participants tend to underestimate the level of cognitive impairment.

5. Conclusions

In summary, use of the time-overlapping intervals strategy enabled us to study the diagnostic instability in MCI subtypes, contributing further knowledge about temporal effects and the time between follow-up assessment on MCI trajectories, and revealing distinctive trajectories of diagnostic changes, especially in the sda-MCI subtype. Nevertheless, the classical prospective approach has some limitations in relation to studying diagnostic instability in MCI. These limitations must be addressed in future studies by using a more comprehensive neuropsychological and biological approaches, longer time intervals between successive follow-ups, and combining prospective and retrospective diagnostic approaches.

Supplementary Materials: The following are available online at http://www.mdpi.com/2076-3425/9/9/242/s1, File S1: The calculations made to study transitions in each time interval.

Author Contributions: Conceptualization, D.F., J.G.-O. and O.J.-R.; Methodology, D.F., J.G.-O., M.P.; Software, J.G.-O., M.P.; Data acquisition, D.F., C.L.-S., O.J.-R. and A.X.P.; Data Processing, C.L.-S. and O.J.-R.; Writing—Original Draft Preparation, D.F.; Writing—Review and Editing, all; Visualization, D.F., J.G.-O. and O.J.-R.; Project Administration, O.J.-R. and A.X.P.; Funding Acquisition, O.J.-R. and A.X.P.

Funding: This work was financially supported through FEDER founds by the Spanish Directorate General of Scientific and Technical Research (Project Ref. PSI2014-55316-C3-1-R), the National Research Agency (Spanish 'Ministry of Science, Innovation and Universities) (Project Ref. PSI2017-89389-C2-1-R) and by the Galician Government (Consellería de Cultura, Educación e Ordenación Universitaria; axudas para a consolidación e estruturación de unidades de investigación competitivas do Sistema Universitario de Galicia; GI-1807-USC: Ref. ED431-2017/27).

Conflicts of Interest: The authors declare no conflict of interest.

References

1. Gerstenecker, A.; Mast, B. Mild cognitive impairment: A history and the state of current diagnostic criteria. *Int. Psychogeriatr.* **2015**, *27*, 199–211. [CrossRef]
2. Peters, K.R.; Katz, S. Voices from the field: Expert reflections on mild cognitive impairment. *Dementia* **2015**, *14*, 285–297. [CrossRef] [PubMed]
3. Petersen, R.C. Mild cognitive impairment as a diagnostic entity. *J. Intern. Med.* **2004**, *256*, 183–194. [CrossRef] [PubMed]
4. Winblad, B.; Palmer, K.; Kiyipelto, M. Mild cognitive impairment–beyond controversies, towards a consensus: Report of the International Working Group on Mild Cognitive Impairment. *J. Intern. Med.* **2004**, *256*, 240–246. [CrossRef] [PubMed]
5. Albert, M.S.; DeKosky, S.T.; Dickson, D.; Dubois, B.; Feldman, H.H.; Fox, N.C.; Gamst, A.; Holtzman, D.M.; Jagust, W.J.; Petersen, R.C.; et al. The diagnosis of mild cognitive impairment due to Alzheimer's disease: Recommendations from the National Institute on Aging and Alzheimer's Association workgroup. *Alzheimer's Dement.* **2011**, *7*, 270–279. [CrossRef]
6. Dubois, B.; Feldman, H.H.; Jacova, C.; DeKosky, S.T.; Barberger-Gateau, P.; Cummings, J.; Delacourte, A.; Galasko, D.; Gauthier, S.; Jicha, G.; et al. Research criteria for the diagnosis of Alzheimer's disease: Revising the NINCDS-ADRDA criteria. *Lancet Neurol.* **2007**, *6*, 734–746. [CrossRef]
7. Petersen, R.C.; Lopez, O.; Armstrong, M.J.; Getchius, T.S.; Ganguli, M.; Gloss, D.; Gronseth, G.S.; Marson, D.; Pringsheim, T.; Day, G.S.; et al. Practice guideline update summary: Mild cognitive impairment: Report of the Guideline Development, Dissemination, and Implementation Subcommittee of the American Academy of Neurology. *Neurology* **2018**, *90*, 126–135. [CrossRef]
8. Díaz-Mardomingo, M.D.C.; García-Herranz, S.; Rodríguez-Fernández, R.; Venero, C.; Peraita, H. Problems in classifying mild cognitive impairment (MCI): One or multiple syndromes. *Brain Sci.* **2017**, *7*, 111. [CrossRef]
9. Facal, D.; Guàrdia-Olmos, J.; Juncos-Rabadán, O. Diagnostic transitions in mild cognitive impairment by use of simple Markov models. *Int. J. Geriatr. Psychiatry* **2015**, *30*, 669–676. [CrossRef]
10. Han, J.W.; Kim, T.H.; Lee, S.B.; Park, J.H.; Lee, J.J.; Huh, Y.; Park, J.E.; Jhoo, J.H.; Lee, D.Y.; Kim, K.W. Predictive validity and diagnostic stability of mild cognitive impairment subtypes. *Alzheimer's Dement.* **2012**, *8*, 553–559. [CrossRef]
11. Cloutier, S.; Chertkow, H.; Kergoat, M.J.; Gauthier, S.; Belleville, S. Patterns of cognitive decline prior to dementia in persons with mild cognitive impairment. *J. Alzheimer's Dis.* **2015**, *47*, 901–913. [CrossRef] [PubMed]
12. Ding, D.; Zhao, Q.; Guo, Q. Progression and predictors of mild cognitive impairment in Chinese elderly: A prospective follow-up in the Shanghai Aging Study. *Alzheimer's Dement.* **2016**, *4*, 28–36. [CrossRef]
13. Klekociuk, S.Z.; Summers, M.J. Exploring the validity of mild cognitive impairment (MCI) subtypes: Multipledomain amnestic MCI is the only identifiable subtype at longitudinal follow-up. *J. Clin. Exp. Neuropsychol.* **2014**, *36*, 290–301. [CrossRef] [PubMed]
14. Cui, Y.; Liu, B.; Luo, S.; Zhen, X.; Fan, M.; Liu, T.; Zhu, W.; Park, M.; Jiang, T.; Jin, J.S.; et al. Identification of conversion from mild cognitive impairment to Alzheimer's disease using multivariate predictors. *PLoS ONE* **2011**, *6*, e21896. [CrossRef] [PubMed]
15. Smith, G.E.; Pankratz, V.S.; Negash, S.; Machulda, M.M.; Petersen, R.C.; Boeve, B.F.; Knopman, D.S.; Lucas, J.A.; Ferman, T.J.; Graff-Radford, N.; et al. A plateau in pre-Alzheimer memory decline: Evidence for compensatory mechanisms? *Neurology* **2007**, *69*, 133–139. [CrossRef] [PubMed]

16. Juncos-Rabadan, O.; Pereiro, A.X.; Facal, D.; Lojo, C.; Caamaño, J.A.; Sueiro, J.; Boveda, J.; Eiroa, P. Prevalence and correlates of mild cognitive impairment in adults aged over 50 years with subjective cognitive complaints in primary care centers. *Geriatr. Gerontol. Int.* **2014**, *14*, 667–673. [CrossRef] [PubMed]

17. Benedet, M.J.; Alejandre, M.A. *TAVEC: Test de Aprendizaje Verbal España-Complutense (TAVEC: Spanish-Complutense Verbal Learning Test)*; TEA Ediciones: Madrid, Spain, 1998.

18. Lobo, A.; Saz, P.; Marcos, G.; Día, J.L.; de la Cámara, C.; Ventura, T.; Morales Asín, F.; Pascual, L.F.; Montañés, J.A.; Aznar, S. Revalidación y normalización del mini-examen cognoscitivo (primera versión en Castellano del mini-mental status examination) en la población general geriatrica. (Revalidation and standardization of the cognition mini-exam (first Spanish version of the mini-mental status examination) in the general geriatric population). *Med. Clin.* **1999**, *112*, 767–774.

19. López-Pousa, S. *CAMDEX-R: Prueba de Exploración Cambridge Revisada Para la Valoración de los Trastornos Mentales en la Vejez. Adaptación Española (CAMDEX-R: Revised Cambridge Screening Test for the Assessment of Mental Disorders in Old Age. Spanish Adaptation)*; TEA Ediciones: Madrid, Spain, 2003.

20. Pauker, J.D. Constructing overlapping cell tables to maximize the clinical usefulness of normative test data: Rationale and an example from neuropsychology. *J. Clin. Psychol.* **1988**, *44*, 930–933. [CrossRef]

21. Langa, K.M.L.; Levine, D.A. The diagnosis and management of mild cognitive impairment: A clinical review. *JAMA* **2014**, *312*, 2551–2562. [CrossRef] [PubMed]

22. Malek-Ahmadi, M. Reversion from mild cognitive impairment to normal cognition. *Alzheimer Dis. Assoc. Dis.* **2016**, *30*, 324–330. [CrossRef] [PubMed]

23. Forlenza, O.V.; Diniz, B.S.; Nunes, P.V.; Memória, C.M.; Yassuda, M.S.; Gattaz, W.F. Diagnostic transitions in mild cognitive impairment subtypes. *Int. Psychoger.* **2009**, *21*, 1088–1095. [CrossRef] [PubMed]

24. Reisberg, B.; Shulman, M.B.; Torossian, C.; Leng, L.; Zhu, W. Outcome over seven years of healthy adults with and without subjective cognitive impairment. *Alzheimer's Dement.* **2010**, *6*, 11–24. [CrossRef] [PubMed]

25. Klekociuk, S.Z.; Summers, J.J.; Vickers, J.C.; Summers, M.J. Reducing false positive diagnoses in mild cognitive impairment: The importance of comprehensive neuropsychological assessment. *Eur. J. Neurol.* **2014**, *21*, 1330–1336. [CrossRef] [PubMed]

26. Edmonds, E.C.; Delano-Woods, L.; Jak, A.J.; Galasko, D.R.; Salmon, D.P.; Bondi, M.W. "Missed" mild cognitive impairment: High false-negative error rate based on conventional diagnostic criteria. *J. Alzheimer's Dis.* **2016**, *52*, 685–691. [CrossRef] [PubMed]

27. Edmonds, E.C.; Delano-Wood, L.; Galasko, D.R.; Salmon, D.P.; Bondi, M.W. Subjective cognitive complaints contribute to misdiagnosis of mild cognitive impairment. *J. Int. Neuropsychol. Soc.* **2014**, *20*, 836–847. [CrossRef] [PubMed]

28. Edmonds, E.C.; Delano-Woods, L.; Clark, L.R.; Jaka, A.J.; Nationd, D.A.; McDonald, C.R.; Libone, D.J.; Auf, R.; Galasko, D.; Salmonh, D.P.; et al. Susceptibility of the conventional criteria for mild cognitive impairment to false-positive diagnostic errors. *Alzheimer's Dement.* **2015**, *11*, 415–424. [CrossRef] [PubMed]

29. Campos-Magdaleno, M.; Facal, D.; Juncos-Rabadán, O.; Picón, E.; Pereiro, A.X. Comparison between an empirically derived and a standard classification of amnestic mild cognitive impairment from a sample of adults with subjective cognitive complaints. *J. Aging Health* **2016**, *28*, 1105–1115. [CrossRef]

MDPI

St. Alban-Anlage 66

4052 Basel

Switzerland

Tel. +41 61 683 77 34

Fax +41 61 302 89 18

www.mdpi.com

Brain Sciences Editorial Office

E-mail: brainsci@mdpi.com

www.mdpi.com/journal/brainsci